磁控溅射制备氮化铜薄膜的结构与性能

肖剑荣 著

中国水利水电出版社
www.waterpub.com.cn
·北京·

内 容 提 要

磁控溅射是一种通用、成熟的薄膜制备工艺技术,其制备工艺可调剂参数较多,通过精细控制能够实现对薄膜结构的有效调控。

本书研究了薄膜的工艺参数、薄膜结构与性能等之间的内在关系,探讨了薄膜的电子输运、光学带隙、热稳定性的有关物理量的变化机制。

全书结构合理,条理清晰,内容丰富新颖,可供相关工程技术人员参考使用。

图书在版编目(CIP)数据

磁控溅射制备氮化铜薄膜的结构与性能/肖剑荣著.
—北京:中国水利水电出版社,2019.1 (2024.10重印)
　ISBN 978-7-5170-7439-7

Ⅰ.①磁… Ⅱ.①肖… Ⅲ.①铜－金属薄膜－氮化处理－结构－研究②铜－金属薄膜－氮化处理－性能－研究 Ⅳ.①TF811

中国版本图书馆 CIP 数据核字(2019)第 031161 号

书　　名	磁控溅射制备氮化铜薄膜的结构与性能 CIKONG JIANSHE ZHIBEI DANHUATONG BAOMO DE JIEGOU YU XINGNENG
作　　者	肖剑荣　著
出版发行	中国水利水电出版社 (北京市海淀区玉渊潭南路1号D座 100038) 网址:www.waterpub.com.cn E-mail:sales@waterpub.com.cn 电话:(010)68367658(营销中心)
经　　售	北京科水图书销售中心(零售) 电话:(010)88383994、63202643、68545874 全国各地新华书店和相关出版物销售网点
排　　版	北京亚吉飞数码科技有限公司
印　　刷	三河市华晨印务有限公司
规　　格	170mm×240mm　16开本　11.25印张　146千字
版　　次	2019年4月第1版　2024年10月第4次印刷
印　　数	0001—2000 册
定　　价	57.00 元

凡购买我社图书,如有缺页、倒页、脱页的,本社营销中心负责调换

版权所有·侵权必究

前 言

科学技术的迅猛发展,各种元器件精密度的提升,对所需材料的要求越来越高,一种新材料的问世对经济的发展将产生巨大的促进作用。薄膜技术及材料已有近 200 年的发展历史,随着薄膜科学技术与薄膜物理学的快速发展,薄膜在微电子、光学、力学等方面应用日益广泛。薄膜产业的日趋壮大和迫切需求又刺激了薄膜技术和薄膜材料的蓬勃发展,在材料领域展现的地位与作用越来越重要,主要包括,宽带隙、发蓝光半导体薄膜材料,高/低介电常数的介质薄膜材料,压电、铁电、巨磁电阻等薄膜材料以及各类耐磨、耐腐、超硬薄膜材料等。这些新型薄膜材料的崛起,为探索材料在低维纳米尺度内的新现象、新规律,材料的新特性、新功能的开发与研制,扩大薄膜材料的应用范围,提高器件的精密性、可靠性和稳定性等提供了有力的保证。薄膜材料不仅展现出优越的电学、光学、力学等方面的优异性能,并且它还可以根据实际需要进行调控,制备出具有某些特定功能的薄膜材料。同时,薄膜材料制备技术相对简单、成本较低和无毒等,因此薄膜材料及技术被视为 21 世纪科学与技术领域的重要发展方向之一。

磁控溅射是一种通用、成熟的薄膜制备工艺技术,其特点是制备的薄膜生长速度快、膜层致密均匀、针孔缺陷少,纯度高,与基体附着好,其制备工艺可调剂参数较多,通过精细控制能够实现对薄膜结构的有效调控。磁控溅射镀膜现已成为真空镀膜的一种通用技术,并广泛应用于科学研究及各行业工业化生产中。本书主要是利用射频/直流磁控溅射技术,在不同的制备参数条件下,制备了氮化铜薄膜(掺杂),研究了薄膜的工艺参数、薄膜结构与性能等之间的内在关系,探讨了薄膜的电子输运、光学带隙、热稳定性的有关物理量的变化机制,希望能为基于低维薄膜材料表面与界面物理性能的研究、相关材料元器件的设计提供参考。

本书是作者及科研团队多年来从事氮化铜薄膜研究工作的总结,是团队共同努力取得的成果,是集体智慧的结晶。在氮化铜薄膜研究工作及本书材料的收集过程中,桂林理工大学理学院材料物理与计算科学专业在读硕士齐孟、郭雅芳、巩晨阳为本书实验样品制备及表征等方面付出了辛勤的劳动;学院教师程勇博士、马家峰博士、王志勇教授等在薄膜样品的测试分析与材料收集等方面给予了大力的支持,在此一并表示感谢。

感谢桂林理工大学化学与生物工程学院李延伟教授、张淑华教授在晶体结构建模与模拟计算方面提出了许多宝贵的意见和建议;感谢地球科学学院张智教授在成书过程中给予了热心的帮助;感谢桂林电子科技大学材料科学与技术学院周昌荣教授在样品测试方面给予精心指导和无私帮助。

本书工作的研究主要是在国家自然科学基金"金属掺杂氮化铜晶体的原位合成及其导电机理研究(No.11364011)"和广西自然科学基金"氮化铜薄膜的调控制备及其低温分解效应机理研究(No.2017GXNSFAA198121)"项目经费支持下完成。

虽然,作者付诸较多的心血致力本书成稿,但是由于学识水平、时间等因素,书中不妥之处在所难免,恳请读者不吝批评指正。

肖剑荣
2018年7月于桂林

目 录

前言

第1章 绪论 ··· 1
    1.1 薄膜简介 ·· 1
    1.2 薄膜的性质 ·· 5
    1.3 薄膜制备技术 ·· 6
    1.4 薄膜的表征技术 ····································· 14

第2章 氮化铜薄膜研究现状 ································ 23
    2.1 Cu_3N 薄膜的制备技术 ······························· 24
    2.2 Cu_3N 薄膜的晶体结构 ······························· 26
    2.3 电学和光学性能 ····································· 29
    2.4 热、力学性能和耐腐蚀性 ····························· 31
    2.5 Cu_3N 薄膜的应用 ·································· 32

第3章 氮化铜薄膜的制备 ·································· 34
    3.1 磁控溅射技术 ······································· 34
    3.2 JGP-450a 型多功能磁控溅射系统 ······················ 41
    3.3 氮化铜薄膜的制备 ··································· 48

第4章 氮化铜薄膜的结构研究 ······························ 51
    4.1 薄膜的结构分析 ····································· 51
    4.2 薄膜的表面形貌 ····································· 57
    4.3 薄膜的组分 ··· 60

  4.4 薄膜的晶格常数 …………………………………… 62

第 5 章 氮化铜的性能研究 ……………………………… 65

  5.1 薄膜的电学性能 …………………………………… 66

  5.2 薄膜的光学性能 …………………………………… 67

  5.3 薄膜的热稳定性研究 ……………………………… 74

第 6 章 氮化铜的第一性原理研究 ……………………… 77

  6.1 概述 ………………………………………………… 77

  6.2 计算方法及过程 …………………………………… 79

  6.3 氮化铜的电子结构计算结果 ……………………… 86

结论与展望 …………………………………………………… 94

参考文献 ……………………………………………………… 97

附录 …………………………………………………………… 111

第 1 章 绪 论

在现今快速发展的社会市场经济中,新材料的价值体现不仅仅是诸多新产品的涌现,更重要的是它广泛渗透于人类的生活,影响着人们的生活质量;它奠定了工业经济与技术的物质基础,成为一个国家经济实力的标志;它推进人类对于自然的新认识,拓展人类的生存能力与发展空间,打造人类对于这个世界新的概念与价值观念。在这个意义上,我们可以说,材料是人类文明的奠基石。材料是支撑工业生产与工业技术的物质基础。在现代社会的经济生活中,诸多高新技术产品都与新材料技术的发展密切相关。新材料技术已经成为一个国家工业水平与技术能力的重要标志。

1.1 薄膜简介

薄膜不同于通常的气态、液态、固态和等离子态,是一种新的凝聚态物质,可为气相、液相和固相,或是它们的组合,也可以是均相的或非均相的,对称的或非对称的,中性的或电荷。从种类上,薄膜可分为金属、半导体和绝缘体,单晶、多晶和非晶态,无机和有机薄膜。随着生产和科学技术的发展,对薄膜的需求日益增长,因而促进了人们对薄膜的研制,特别是近年来薄膜的研制工作进展很快。薄膜材料应用非常广泛,主要应用在包装材料方面,随着新材料的兴起,现已经渗透至各个领域。近年来,薄膜在电工电子、太阳能材料方面得到快速的发展,薄膜在各领域的应用比例如图 1-1 所示。

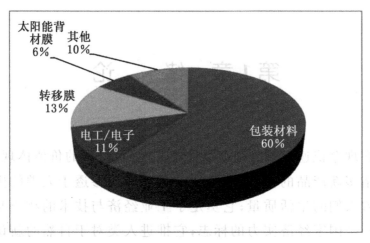

图 1-1 薄膜各领域应用情况

薄膜及相关薄膜器件兴起于 20 世纪 60 年代，是新理论与高技术高度结晶的产物，已成为电子、信息、传感器、光学、太阳能利用等技术的核心基础，在卷镀薄膜产品、塑料金属化制品、建筑镀膜制品、光学薄膜、集成电路、太阳能电池、液晶显示膜、刀具硬化膜、光盘、磁盘等方面都已具有相当大的生产和市场规模，并且与材料科学和工程中的其他学科相互渗透、相互促进，是 21 世纪材料科学与工程的关键所在，更是世界各国高技术竞争的焦点。发展薄膜的基础与应用研究，真正的意义是为了加快高技术应用的步伐，加强传统学科知识和现代学科知识的有机联系，不仅可以扩充块体材料的应用领域与利用高性能的薄膜器件，而且可以通过薄膜的制备技术和自身的特殊结构形式，获得新的人工合成物质与发现新的物理化学性质，创新科学理论。

目前，对薄膜材料的研究正在向多种类、高性能、新工艺等方面发展，其基础研究也在向分子层次、原子层次、纳米尺度、介观结构等方向深入，新型薄膜材料的应用范围正在不断扩大。当前薄膜科学与技术得到迅猛发展的主要原因是，新型薄膜材料的研究工作始终同现代高新技术相联系，并得到广泛的应用，常用的有超导薄膜、导电薄膜、电阻薄膜、半导体薄膜、介质薄膜、绝缘薄膜、钝化与保护薄膜、压电薄膜、铁电薄膜、光电薄膜、磁电薄膜、磁光薄膜等。近 10 年来,新型薄膜材料在以下几个方面的发展更为突出:

(1) 金刚石薄膜。金刚石薄膜的禁带宽,电阻率和热导率大,载流子迁移率高,介电常数小,击穿电压高,是一种性能优异的电子薄膜功能材料,应用前景十分广阔。金刚石薄膜有很多优异的性质:硬度高、耐磨性好、摩擦系数高、化学稳定性好、热导率高、热膨胀系数小,是优良的绝缘体。金刚石薄膜属于立方晶系,面心立方晶胞,每个晶胞含有 8 个 C 原子,每个 C 原子采取 sp^3 杂化与周围 4 个 C 原子形成共价键,牢固的共价键和空间网状结构是金刚石硬度很高的原因。利用它的高导热率,可将它直接积在硅材料上成为既散热又绝缘的薄层,是高频微波器件、超大规模集成电路最理想的散热材料。利用它的电阻率大,可以制成高温工作的二极管,微波振荡器件和耐高温高压的晶体管以及毫米波功率器件等。

(2) 铁电薄膜。铁电薄膜的制备技术和半导体集成技术的快速发展,推动了铁电薄膜及其集成器件的实用化。铁电材料已经应用于铁电动态随机存储器(FDRAM)、铁电场效应晶体管(FEET)、铁电随机存储器(FFRAM)、IC 卡、红外探测与成像器件、超声与声表面波器件以及光电子器件等十分广阔的领域。铁电薄膜的制作方法一般采用溶胶-凝胶法、离子束溅射法、磁控溅射法、有机金属化学蒸汽沉积法、准分子激光烧蚀技术等。已经制成的晶态薄膜有铌酸锂、铌酸钾、钛酸铅、钛酸钡、钛酸锶、氧化铌和锆钛酸铅等,以及大量的铁电陶瓷薄膜材料。

(3) 半导体复合薄膜。以非晶硅氢合金薄膜(a—Si:H)和非晶硅基化物薄膜(a—SiGe:H、a—SiC:H、a—SiN:H 等)为代表。它有良好的光电特性,可以应用于太阳能电池,其特点是:廉价、高效率和大面积化。为了改善这些器件的性能,又研制了多晶硅膜、微晶硅膜及纳米晶硅薄膜。这些器件已列入各国发展计划中,如日本的"阳光计划",欧洲的"焦耳-热量计划",美国的"百万屋顶计划",中国的"973"和"863"计划,并已发展成为高新技术产业,另一项有发展前途的是 Cu(InGa)Se$_2$(小面积效率＞18.8%)及 η 为 16.4% 的 CdTe 薄膜太阳电池也列入国家"863"计划。这类半导体薄膜复合材料,特别是硅薄膜复合材料已开始用于低功耗、低噪声的大规模集成电路中,以减小误差,提高电路的抗辐射能力。

(4)超晶格薄膜材料。随着半导体薄膜层制备技术的提高,当前半导体超晶格材料的种类已由原来的砷化镓、镓铝砷扩展到铟砷、镓锑、铟铝砷、铟镓砷、碲镉、碲汞、锑铁、锑锡碲等多种。组成材料的种类也由半导体扩展到锗、硅等元素半导体,特别是今年来发展起来的硅、锗硅应变超晶格,由于它可与当前硅的平面工艺相容和集成,格外受到重视,甚至被誉为新一代硅材料。

半导体超晶格结构不仅给材料物理带来了新面貌,而且促进了新一代半导体器件的产生,除上面提到的可制备高电子迁移率晶体管、高效激光器、红外探测器外,还能制备调制掺杂的场效应管、先进的雪崩型光电探测器和实空间的电子转移器件,并正在设计微分负阻效应器件、隧道热电子效应器件等,它们将被广泛应用于雷达、电子对抗、空间技术等领域。

(5)纳米复合薄膜。随着纳米材料的出现,纳米薄膜(涂层)技术也得到相应的发展。时至今日,已从单一材料的纳米薄膜转向纳米复合薄膜的研究,薄膜的厚度也由数微米发展到数纳米的超薄膜。纳米复合薄膜是指由特征维度尺寸为纳米数量级(1～100nm)的组元镶嵌于不同的基体里所形成的复合薄膜材料,有时也把不同组元构成的多层膜如超晶格称为纳米复合薄膜,它具有传统复合材料和现代纳米材料两者的优越性。

纳米复合薄膜是一类具有广泛应用前景的纳米材料,按用途可分为两大类,即纳米复合功能薄膜和纳米复合结构薄膜。前者主要利用纳米粒子所具有的光、电、磁方面的特异性能,通过复合赋予基体所不具备的性能,从而获得传统薄膜所没有的功能。而后者主要通过纳米粒子复合提高机械方面的性能。由于纳米粒子的组成、性能、工艺条件等参量的变化都对复合薄膜的特性有显著的影响,因此可以在较多自由度的情况下人为地控制纳米复合薄膜的特性。组成复合薄膜的纳米粒子可以是金属、半导体、绝缘体、有机高分子等材料,而复合薄膜的基体材料可以是不同于纳米粒子的任何材料。人们采用各种物理和化学方法先后制备了一系列金属/绝缘体、半导体/绝缘体、金属/半导体、金属/高分子、半导体/高分子等纳米复合薄膜。特别是硅系纳米复合薄膜材料得到了深入的研究,人们利用热蒸发、溅射、等离子体气相沉积等各种方法制备了 Si/SiO_x、Si/a—$Si:H$、Si/SiN_x、Si/SiC 等纳米镶嵌复合薄膜。尽管目前对其机制不十分清楚,却通过大量实验现象发现在此类纳米复合薄膜中观察到了强的从红外到紫外的可见光发射。由于这一类薄膜稳定性大大高于多孔硅,工艺上又可与集成电路兼容,因而被期待作为新型的光电材料应用于大规模光电集成电路。

由于纳米复合薄膜的纳米相粒子的量子尺寸效应、小尺寸效应、表面效应、宏观量子隧道效应等使得它们的光学性能、电学性能、力学性能、催化性能、生物性能等方面呈现出常规材料不具备的特性。因此,纳米复合薄膜在光电技术、生物技术、能源技术等各个领域都有广泛的应用前景。现以硅系纳米复合薄膜材料为例介绍它们的特性及其应用。

1.2 薄膜的性质

薄膜是由离子、原子或分子的沉积过程形成的二维材料。很难给薄膜下一般性的定义,可以这样理解:采用一定方法,使处于某种状态的一种或几种物质(原材料)的基团以物理或者化学方式附着于某种物质(衬底材料)表面,在衬底材料表面形成一层新的物质,这层新的物质就称为薄膜。

薄膜的基本性质是:具有二维延展性,其厚度方向的尺寸远远小于其他两个方向的尺寸。不管是否能够形成自持(自支撑)的薄膜,衬底材料是必备的前提条件,即只有在衬底表面才能获得薄膜。广义上,薄膜包括气态、液态和固体三种形态,分别称为气态薄膜、液态薄膜和固体薄膜。这里,我们所指的薄膜仅仅只是固体薄膜。按结晶状态,薄膜可以分为非晶态与晶态,后者进一步分为单晶薄膜和多晶薄膜。单晶薄膜的概念是从外延生长,特别是同质外延和结晶学而来,在单晶衬底材料上进行同质或异质外延,要求外延薄膜连续、平滑且与衬底材料的晶体结构存在对应关系,并且是一种定向生长。这就要求单晶薄膜不仅在厚度方向有晶格的连续性,而且在衬底材料表面方向也有连续性。也有人将取向生长或织构生长的薄膜称为单晶薄膜,但没有严格地考虑晶界或界面。多晶薄膜是指在一个衬底材料上生长的由许多取向相异单晶集合体组成的薄膜。相对于晶态薄膜,非晶态薄膜是指在薄膜结构中原子的空间排列表现为短程有序和长程无序。从化学角度,薄膜可以分为有机薄膜和无机薄膜。依组成元素,可以分为金属薄膜与非金属薄膜。按物理性能,可以划分为硬质薄膜、声学薄膜、热学薄膜、金属导电薄膜、半导体薄膜、超导薄膜、介电薄膜、磁阻薄膜、光学薄膜等。与块体材料相比,由于薄膜厚度方向的尺寸很小,显示出明显的尺寸效应,表现出一些块体材料不具备的力、声、热、电、光等物理特性。

1.3 薄膜制备技术

制备薄膜的方法较多,从物理作用和化学反应角度,也就是按学科分为:物理成膜和化学成膜方法;代表性的薄膜制备方法有:物理气相沉积(PVD)、化学气相沉积(CVD)等。

通常,薄膜制备中需考虑的主要问题有:

(1)制备方法的选择与制备技术的提高。

(2)工艺流程的优化及其与平面工艺的兼容性。

(3) 降低制备成本与提高薄膜器件性能之间的平衡。

(4) 制备过程的安全性及其对环境的影响。

1.3.1 物理气相沉积

一般说来,物理气相沉积是把固态或液态成膜材料通过某种物理方式(如高温蒸发、溅射、等离子体、离子束、激光束等)产生气相原子、分子、离子,再经过输运在基体表面沉积,或与其他活性气体反应形成产物在基体上沉积为固相薄膜的过程。物理气相沉积需用固态或熔化态的物质作为沉积过程的源物质,源物质需经过物理过程转变为气相,工作环境需要较低的气压,在气相中和衬底表面一般不发生化学反应,但反应沉积例外。

1. 真空蒸发镀膜

在真空室内加热,使固态原材料蒸发汽化或升华,并凝结沉积到一定温度的衬底的表面,形成薄膜,这就是真空蒸发镀膜技术,简称蒸发镀,它是一种非常简单的薄膜制备技术,真空蒸镀工作示意图如图1-2所示。真空蒸发镀膜分为三个基本过程:

(1) 被蒸发材料的加热蒸发过程:通过一定加热方式,使被蒸发材料受热蒸发。

(2) 气态原子或分子由蒸发源到衬底的输运过程:该过程蒸发或升华,即由固态或液态转变为气态;主要受真空度、蒸发源-衬底间距、被蒸发材料蒸气压的影响。

(3) 衬底表面的沉积过程:包括粒子与衬底表面的碰撞、粒子在衬底表面的吸附与解吸、表面迁移以及成核和生长等过程。根据蒸发源加热方式不同,可以分为几种不同的真空蒸发镀膜方法。常见加热方式有电阻加热、电子束加热、高频感应加热、电弧加热和激光加热等。

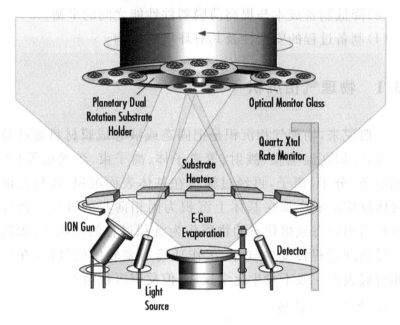

图 1-2 真空蒸镀工作示意图

2. 溅射镀膜

溅射是指利用气体放电产生的正离子,在电场作用下加速成为高能粒子,撞击固体(靶)表面,进行能量和动量交换后,固体表面的原子或分子在轰击下离开表面,利用固体表面被溅射出来的物质沉积成膜的过程,称溅射镀膜。它与真空蒸发镀膜的区别是:一个以动量转换为主,一个以能量转换为主。

溅射镀膜技术的制膜范围较宽,可以用来制备金属膜、导体膜、氧化物膜等。溅射镀膜法较其他镀膜有很多优点:

(1)镀膜过程中无相变现象,使用的薄膜材料非常广泛。

(2)沉积粒子能量大,并对衬底有清洗作用,薄膜附着性好。

(3)薄膜密度高,杂质少。

(4)膜厚可控性、重复性好。

(5)可以制备大面积薄膜;但溅射镀膜也有不足之处,如设备复杂,需要高压,沉积速率低。

入射荷能离子轰击靶材表面会发生一系列物理、化学现象。

它包括以下三类现象：

(1) 靶材表面产生原子或分子溅射，二次电子发射，正、负离子发射，溅射原子返回，杂质原子解吸附或分解，光子辐射等。

(2) 产生表面物理、化学现象，如加热、清洗、刻蚀、化学分解或反应。

(3) 材料表面层发生结构损伤、碰撞级联、离子注入、扩散共混、非晶化等现象。其实，物体置于等离子体中，其表面具有一定的负电位时，就会发生溅射现象，只需要调整其相对等离子体的电位，就可以获得不同程度的溅射效应，从而实现溅射镀膜，溅射清洗或溅射刻蚀及辅助沉积过程。

溅射的基本工作原理是利用辉光放电时正离子对阴极溅射，当作用于低压气体的电场强度超过某临界值时，将出现气体放电现象。气体放电时在放电空间产生大量电子和正离子，在极间电场作用下迁移运动形成电流。根据溅射设备的不同，溅射镀膜可以分为直流溅射、射频溅射和磁控溅射等几种不同的镀膜方式。

3. 脉冲激光沉积

脉冲激光沉积法（PLD 法）是 20 世纪 80 年代后期发展起来的一种物理沉积方法，该法是很有竞争力的新工艺。该法利用激光对物体进行轰击，然后将轰击出来的物质沉淀在不同的衬底上，得到沉淀或者薄膜，其工作示意图如图 1-3 所示。

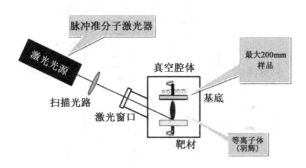

图 1-3 脉冲激光沉积工作示意图

PLD 作为一种新的先进的成膜技术。与其他工艺相比，生长参数独立可调、可精确控制化学计量比,易于实现超薄薄膜的生长和多层膜的制备,生长的薄膜结晶性能很好,膜的平整度也较高。PLD 技术的成膜效率高,能够进行批量生产,这是其很大的优势,有望在高质量 ZnO 薄膜的研究和生产中得到广泛的应用。但是由于等离子体管中的微粒、气态原子和分子沉积在薄膜上会降低薄膜的质量,采取相应的措施后可以获得改善,但不能完全消除。而且 PLD 生长在控制掺杂、生长平滑的多层膜和厚度均匀等方面都比较困难,从而比较难以进一步提高薄膜的质量。

4. 离子镀

溅射法是利用被加速的正离子的撞击作用,使蒸气压降低而难蒸发的物质变成气体,这种正离子若打到基片上,还会起到表面清洁的作用,提高薄膜质量,然而这导致了成膜速度受到一定限制。为了解决这一问题,将真空镀膜与溅射镀膜结合起来,利用真空蒸镀来镀膜,利用溅射来清洗基片表面,这种制膜方法被称为离子镀膜。离子镀膜是利用电弧和热丝电子产生沉积原子或离子,然后通过电磁场加速并沉积在基底表面,形成薄膜,其工作示意图如图 1-4 所示。它是结合真空蒸发镀膜和溅射镀膜的特点而发展起来的一种新型镀膜技术。该技术是采用将蒸发或溅

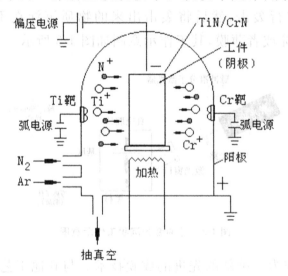

图 1-4 离子镀工作示意图

射与离子注入相结合的方法,具有沉积温度低,薄膜生长致密和成本低廉等优点。但该技术的缺陷在于薄膜的沉积速率低,不能在大的基体上沉积薄膜,不适合大规模工业化生产。

1.3.2 化学气相沉积

化学气相沉积是一种化学气相生长法,简称 CVD(Chemical Vapor Deposition)技术,它是利用流经基片表面的气态物料,借助加热、等离子体、紫外光或激光等作用而在基片表面发生化学反应,形成薄膜的一种方法。由于 CVD 是一种化学反应方法,利用这一技术可以在各种基片上制备多种薄膜,如各种单晶、多晶、非晶、单相或多相薄膜。

CVD 与 PVD 的区别就在于 CVD 依赖于化学反应生成固态薄膜,相对于其他薄膜沉积技术具有许多优点:

(1)设备、操作简单,可通过气体原料流量的调节,在较大范围控制产物组分,可制备梯度膜、多层单晶膜及多层膜的微组装。

(2)薄膜晶体质量好,薄膜致密,膜层纯度高,适用于金属、非金属及合金等多种膜的制备。

(3)可在远低于熔点或分解温度下实现难熔物的沉积,且薄膜黏附性好。

(4)反应所需原材料易获得。

(5)可进行可控杂质掺杂。

(6)可获得平滑沉积表面,且易实现外延。

(7)可在常压和低真空下进行。

CVD 的主要缺点是,需要在高温下反应,基片温度高(一般在 1000℃),沉积速率较低,一般每小时只有几微米到几百微米,使用的设备较电镀法复杂,基体难于进行局部沉积,以及参加沉积反应的气源和反应后的余气都有一定毒性等,因此它的应用不如蒸发镀膜、溅射镀膜那样广泛。

1.3.3 其他镀膜技术

1. 分子束外延

分子束外延(MBE)是在真空蒸发技术基础上改进而来的。在超高真空下,将各组成元素的分子束流,以一个个分子的形式喷射到衬底表面,在适当的衬底温度等条件下外延沉积。其优点是可以生长极薄的单晶层,可用于制备超晶格、量子点等,在固态微波器件、光点器件、超大规模集成电路等领域广泛应用。特点为:超高真空下生长,薄膜所受污染小;生长过程和生长速率严格可控;膜的组分和掺杂浓度可通过源的变化迅速调整;生长速率低,可实现单原子层的控制生长;衬底温度低,有利于减小自掺杂;衬底与束能源分开,有利于生长过程、表面成分、晶体结构的实时观察,研究生长机制;能有效利用平面技术。

2. 化学溶液镀膜

化学溶液镀膜法是指在溶液中利用化学反应或电化学原理在基体材料表面上沉积成膜的一种技术。它包括各种化学反应沉积、阳极氧化、电镀等。化学溶液镀膜的特点是:

(1)可在复杂的镀件表面形成均匀的镀层。

(2)不需要导电电极。

(3)通过敏化处理活化,可直接在塑料、陶瓷、玻璃等非导体上镀膜。

(4)镀层孔隙率低。

(5)镀层具特殊的物理、化学性质。

目前已广泛用于镀镍、钴、铂、铜、银、金等金属膜,并为非金属材料镀覆金属膜层开辟了广阔的前景。

3. 电化学沉积

电化学沉积主要包括电镀和阳极氧化两种方法。电镀是利用电解反应,在处于负极的衬底上进行镀膜的方法,又称湿式镀膜技术。电镀中所用的电解液称为电镀液或镀液,一般用来镀金属的盐类,可为单盐和络盐两类,含单盐的镀液如硫酸盐、氯化物等,含络盐的镀液如氰化物等。单盐使用安全、价格便宜,但所得的膜层较粗糙;络盐价格贵、毒性大,但容易得到致密的膜。使用时可根据不同的要求,选择不同的镀液。对镀层的基本要求是:具有细致紧密的结晶,镀层平整,光滑牢固,无针孔麻点等。要获得良好的镀层,金属零件在镀前必须进行彻底清洗,否则很难镀上,或镀层容易起泡剥落。由于电镀在常温下进行,所得的膜层具有细致紧密、平整、光滑牢固、无针孔、不粗糙等优点,并且厚度容易控制,因而它在电子工业中得到了广泛的应用。

4. LB 膜技术

LB 膜技术首先由 Langmuir 和 Blodett 使用,称 Langmuir-Blodett 法,简称 LB 法。这种技术是 20 世纪二三十年代由美国科学家 Langmuir 及其学生 Katharine Blodgett 建立的一种单分子膜制备技术,它是将兼具亲水头和疏水尾的两亲性分子分散在水面上,经逐渐压缩其水面上的占有面积,使其排列成单分子层,再将其转移沉积到固体基底上所得到的一种膜。

LB 膜技术优点:

(1)膜厚为分子级水平(纳米数量级),具有特殊的物理化学性质。

(2)可以制备单分子膜,也可以逐层累积形成多层 LB 膜,组装方式任意选择。

(3)可以人为选择不同的高分子材料,累积不同的分子层,使之具有多种功能。

(4)成膜可在常温常压下进行,所需能量小,基本不破坏成膜材料的结构。

(5)控制膜层厚度及均匀性方面远比常规制膜技术优越。

(6)可有效地利用 LB 膜分子自身的组织能力,形成新的化合物。

(7)LB 膜结构容易测定,易于获得分子水平上的结构与性能之间的关系。

LB膜技术缺点：

（1）由于LB膜淀积在基片上时的附着力是依靠分子间作用力，属于物理键力，因此膜的机械性能较差。

（2）要获得排列整齐而且有序的LB膜，必须使材料含有两性基团，这在一定程度上给LB成膜材料的设计带来困难。

（3）制膜过程中需要使用氯仿等有毒的有机溶剂，这对人体健康和环境具有很大的危害性。

（4）制膜设备昂贵，制膜技术要求很高。

1.4　薄膜的表征技术

随着科学技术的发展，薄膜的测试分析技术也日臻完善，表征技术也很多，本书简单介绍应用于氮化铜薄膜的表征技术。

1.4.1　薄膜的结构表征

薄膜结构的研究可以根据其尺寸范围分为三个层次：薄膜的宏观形貌，包括薄膜的尺寸、形状、厚度、均匀性等；薄膜微观形貌，如晶粒及物相的尺寸大小和分布、孔洞和裂纹、界面扩散层及薄膜结构等；薄膜的显微组织，包括晶粒内的缺陷、界面的完整性等。根据需要可选择不同的研究手段：光学金相显微镜、扫描电子显微镜、透射电子显微镜、原子力显微镜以及X射线衍射技术等，在本书中还采用喇曼光谱对薄膜的结构形态进行了分析。

1. 扫描电子显微镜

扫描电子显微镜（SEM）的主要工作模式之一就是二次电子模式。二次电子是入射电子从样品表层激发出来的能量较低的一部分电子，二次电子的能量低，说明它来自样品表面最外层的几个原子。由于样品表面的起伏变化将造成二次电子发射的数量及角度分布的变化,通过保持屏幕扫描与样品表面电子扫描的同步,即可以使屏幕图像重现样品的表面形貌,而屏幕上图像的大小与实际样品上的扫描面积大小之比即是扫描电子显微镜的放大倍数。其工作原理及成像原理图,如图1-5所示。为了防止样品表面的电荷积累而影响观察,样品必须具有导电能力,导电能力差的样品可以喷涂一层导电性较好的C或Au膜的方法。

采用JSF-2100型号扫描电子显微镜测试了单晶硅上的Cu_3N薄膜样品表面形貌和断面形貌,并获得了其表面能谱。

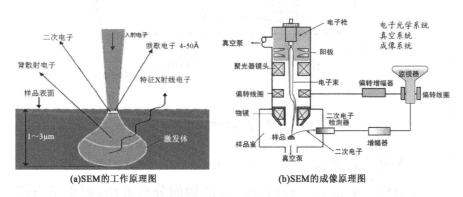

图1-5 SEM工作原理及成像原理图

2. 原子力显微镜

原子力显微镜(AFM)是扫描探针显微镜中的一种,它测量的是物质原子间的作用力。原子力显微镜的基本原理是:将一个对微弱力极敏感的微悬臂一端固定,另一端有一微小的针尖,针尖与样品表面轻轻接触,由于针尖尖端原子与样品表面原子间存在极微弱的排斥力,通过在扫描时控制这种力的恒定,带有针尖的微悬臂将对应于针尖与样品表面原子间作用力的等位面而在垂直于样品的表面方向起伏运动。利用光学检测法或隧道电流检测法,可测得微悬臂对应于扫描各点的位置变化。当原子间的距离减小到一定程度以后,原子间的作用力将迅速上升。因此,由显微探针受力的大小就可以直接换算出样品表面的高度,从而获得样品表面形貌的信息。原子力显微镜工作原理图,如图1-6所示。

本书采用 Solver P47 型原子力显微镜观测了单晶硅上 Cu_3N 薄膜样品的表面形貌。

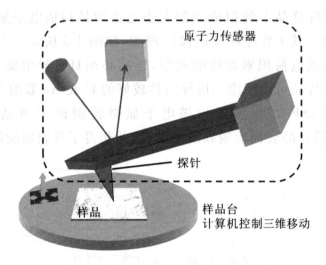

图 1-6 原子力显微镜工作原理图

3. X 射线衍射法

X 射线衍射法是一种研究晶体结构的分析方法,而不是直接研究试样内含有元素的种类及含量的方法。当 X 射线照射晶态结构时,将受到晶体点阵排列的不同原子或分子所衍射。X 射线照射两个晶面距为 d 的晶面时,受到晶面的反射,两束反射 X 射线光程差 $2d\sin\theta$ 是入射波长的整数倍时,即 $2d\sin\theta=n\lambda$(n 为整数),两束光的相位一致,发生相长干涉,这种干涉现象称为衍射,晶体对 X 射线的这种折射规则称为布拉格规则。θ 称为衍射角(入射或衍射 X 射线与晶面间夹角)。n 相当于相干波之间的位相差,$n=1,2\cdots$时各称 0 级、1 级、2 级……衍射线。反射级次不清楚时,均以 $n=1$ 求 d。晶面间距一般为物质的特有参数,对一个物质若能测定数个 d 及与其相对应的衍射线的相对强度,则能对物质进行鉴定,X 射线衍射工作原理及构成如图 1-7 所示。由于原子在空间呈周期性排列,因而这些散射只能在某些方向叠加而产生干涉现象,形成衍射峰。对于非晶态固体,原子在空间是无规则排列的,所以,没有衍射特征峰。但短程有序的存在使得在低角度衍射范围仍具有择优性的衍射极大,形成非晶态谷包。由这些非晶态谷包的分布强度,运用傅里叶变换不但可以获得无规则网络结构的临近关系,还可以检验无规则网络的有序度。

本书中用 X'pert3 Powder 型 X 射线衍射仪(XRD)测试了单晶硅上 Cu_3N 薄膜的晶体结构,以 Cu 靶 $K\alpha$ 射线作为衍射源,X 光波长为 0.15406nm。

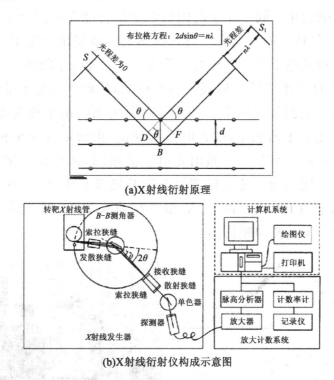

图 1-7　X 射线衍射原理及构成示意图

4. X 射线光电子能谱

X 射线光电子能谱技术(X-ray Photoelectron Spectroscopy,XPS)是电子材料与元器件显微分析中的一种先进分析技术,而且是和俄歇电子能谱技术(AES)常常配合使用的分析技术。由于它可以比俄歇电子能谱技术更准确地测量原子的内层电子束缚能及其化学位移,所以它不但为化学研究提供分子结构和原子价态方面的信息,还能为电子材料研究提供各种化合物的元素组成和含量、化学状态、分子结构、化学键方面的信息。它在分析电子材料时,不但可提供总体方面的化学信息,还能给出表面、微小区域和深度分布方面的信息。另外,因为入射到样品表面的X射线束是一种光子束,所以对样品的破坏性非常小。

X射线光子的能量为 1000~1500eV,不仅可使分子的价电子电离而且也可以把内层电子激发出来,内层电子的能级受分子环境的影响很小。同一原子的内层电子结合能在不同分子中相差很小,故它是特征的。光子入射到固体表面激发出光电子,利用能量分析器对光电子进行分析的实验技术称为光电子能谱。

XPS的原理是用X射线去辐射样品,使原子或分子的内层电子或价电子受激发射出来。被光子激发出来的电子称为光电子。可以测量光电子的能量,以光电子的动能/束缚能为横坐标,相对强度(脉冲/s)为纵坐标可做出光电子能谱图,从而获得试样有关信息。X射线光电子能谱原理示意图及仪器照片,如图1-8所示。

本书采用ESCALAB250Xi型全自动X光电子能谱仪测量了单晶硅上的Cu_3N薄膜样品。

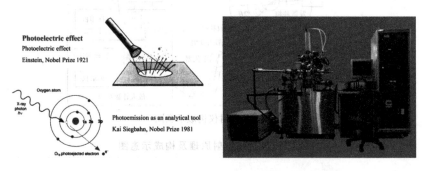

图 1-8 X射线光电子能谱原理示意图及仪器照片

1.4.2 薄膜的性能表征

1. 紫外—可见光吸收光谱

近紫外区及可见光区(200~700nm)是研究光谱最方便的区域,一般研究分子的吸收光谱,常用的仪器是紫外—可见光分光光度计。紫外-可见光分光光度计基本装置中采用高强度的钨灯作为光源,能够在可见光范围(400~700nm)调节。氘灯用于紫外分光光度测量(200~400nm);使用氘灯时要用石英杯,因为紫外线不能透过玻璃。分光光度计的优点就在于使用了一个衍射光栅将光源的复色光转换为单色平行光束。分子的紫外及可见吸收光谱是由于电子能级的跃迁产生的(伴随有振动、转动能级的改变)。电子的能级跃迁主要是价电子的跃迁,因为内部电子的能级很低,在一般情况下不易激发。

采用TU1800型紫外/可见光分光光度计测量了沉积在石英片上的Cu_3N薄膜的透射光谱,为消除石英基片所带来的影响,在进行测试时,用一片未镀膜的石英基片作为校准,光波扫描的范围在200~800nm。

2. 椭偏仪测试

椭圆偏振光法是一种可以同时测定固体薄膜的厚度及其折射率的有效方法,其精密度在0.1nm以下。与其他常用测量方法相比,其测量精密度高、可测范围宽,厚度从0.1nm到$10\mu m$的透明薄膜都可以用它进行测量,并且用该方法测量,对样品的破坏性小。其原理是:当一束光以一定的入射角照射透明薄膜时,入射光要在薄膜的前后界面发生多次反射和折射,反射光束的振幅和位相的变化与薄膜的厚度及折射率有关;如果入射光是椭圆偏振光,则只要测量反射光偏振状态的变化,就可以确定薄膜的厚度和折射率。

采用ELLI-B型椭圆偏振光测试仪测试了单晶硅上Cu_3N薄膜样品的厚度和折射率。

3. 四探针测试仪

本书采用SZT-90型四探针测试仪获得了Cu_3N薄膜的电阻率。SZT-90型数字式四探针测试仪是运用四探针测量原理的综合测量仪,适用于对半导体、金属、绝缘体材料的电阻性能测试,可以测量块状半导体材料的径向和轴向电阻率,片状(薄膜)半导体材料的电阻率和扩散层方块电阻。仪器具有测量精度高、灵敏度高、稳定性好,使用方便的特点,测量结果由数字直接显示。

直流四探针法测试原理简介:

当1、2、3、4根金属探针排成直线时,并以一定的压力压在半导体材料上,在1、4两处探针间通过电流I,则2、3探针间产生电位差V,如图1-9所示。

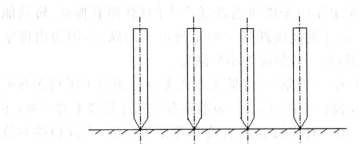

图1-9 四探针测试工作简图

薄片样品因为其厚度与探针间距比较,不能忽略,测量时要提供样品的厚度、形状和测量位的修正系数。

电阻率可由下面公式得出:

$$\rho = 2\pi S \frac{V}{I} G\left(\frac{W}{S}\right) D\left(\frac{d}{s}\right) = \rho_0 G\left(\frac{W}{S}\right) D\left(\frac{d}{s}\right)$$

式中,ρ_0为块形体电阻率测量值;$G\left(\frac{W}{S}\right)$为样品厚度与探针间距的修正函数,可由相关表格查得;$D\left(\frac{d}{s}\right)$为样品形状和测量位置的修正函数。

当圆形硅片的厚度满足$W/S<0.5$时,电阻率为:

$$\rho = \rho_0 \left(\frac{W}{S}\right) D\left(\frac{d}{s}\right) = \frac{\pi}{\ln 2} \frac{V \cdot M}{I} d\left(\frac{d}{s}\right) = 4.53 \frac{V}{L} W D\left(\frac{d}{s}\right)$$

4. 热重分析

在加热或冷却物质的过程中,随着物质的结构、相态和化学性质的变化,通常伴有相应的物理性质的变化,包括质量、温度、热量以及机械声学、电学、光学、磁学等性质,依此构成了相应的热分析测试技术。热分析是测量材料的性质随温度的变化,它在表征材料的热性能、物理性能、机械性能以及稳定性等方面有着广泛的应用,对于材料的研究开发和生产中的质量控制具有很重要的实际意义。

热重法是在程序控制温度下,测量物质的质量与温度关系的一种技术。热重法记录的是热重曲线(TG 曲线),它是以质量作纵坐标,从上向下表示质量减少;以温度(T)或时间(t)作横坐标,自左向右表示增加。只要物质受热时发生质量的变化,就可用热重法来研究其变化过程。TG 主要用于材料的组成分析、热稳定性评价、氧化或分解反应及其动力学研究、材料老化研究等。

采用德国 NETZSCH TG 209C 型热重分析仪对薄膜样品进行了热分析,仪器天平精确至 0.001mg,加热过程中样品通高纯氮气进行了保护,加热温度 30～700℃,幅度 10℃/min。

5. 荧光磷光光度计

在材料物理性能的研究中,发光光谱是研究固体中电子状态、电子跃迁过程以及电子－晶格相互作用等物理问题的一种常用方法。通过固体粉末材料－电子俘获材料荧光光谱的测定,了解固体荧光产生的机理和一些相关的概念,学习荧光光谱仪的结构和工作原理,掌握荧光光谱的测量方法,并对荧光光谱在物质特性分析和生产实际中的应用有初步的了解。

用于测定荧光光谱的仪器称为荧光分光光度计。荧光分光光度计的主要部件有:激发光源、激发单色器(置于样品池后)、发射单色器(置于样品池后)、样品池及检测系统。荧光分光光度计一般采用氙灯作光源,氙灯所发射的谱线强度大,而且是连续光谱、连续分布在 250～700nm 波长范围内,并且在 300～400nm 波长之间的谱线强度几乎相等。

激发光经激发单色器分光后照射到样品室中的被测物质上,物质发射的荧光再经发射单色器分光后经光电倍增管检测,光电倍增管检测的信号经放大处理后送入计算机的数据采集处理系统从而得到所测的光谱。计算机除具有数据采集和处理的功能外,还具有控制光源、单色器及检测器协调工作的功能。

用 Cary Eclipse 型荧光磷光光度计研究了石英片上薄膜的光致发光特性,采用 370nm 和 310nm 波长的 He-Cd 激光。

6. 霍尔效应测试仪

霍尔效应是导电材料中的电流与磁场相互作用而产生电动势的效应。1879 年美国霍普金斯大学研究生霍尔在研究金属导电机构时发现了这种电磁现象,故称霍尔效应。后来曾有人利用霍尔效应制成测量磁场的磁传感器,但因金属的霍尔效应太弱而未能得到实际应用。随着半导体材料和制造工艺的发展,人们又利用半导体材料制成霍尔元件,由于它的霍尔效应显著而得到实用和发展,现在广泛用于非电量检测、电动控制、电磁测量和计算装置方面。在电流体中的霍尔效应也是目前在研究中的"磁流体发电"的理论基础。近年来,霍尔效应实验不断有新发现。1980 年西德物理学家冯·克利青(K. Von Klitzing)研究二维电子系统的输运特性,在低温和强磁场下发现了量子霍尔效应,是凝聚态物理领域最重要的发现之一。

本书利用 HMS-5300 型霍尔效应测试仪测试了氮化铜薄膜的电学性能。

第 2 章 氮化铜薄膜研究现状

氮化铜(Cu_3N)薄膜具有特殊结构和物理化学特性,近年来受到了众多研究者的高度重视,它已成为了半导体材料的一个研究热点[1-18]。图 2-1 所示是 1999—2018 年 Web of Science(SCI)引文数据库中有关 Cu_3N 研究论文的发表数量及论文被引数量的

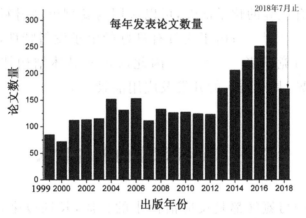

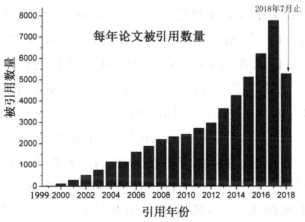

图 2-1 1999—2018 年 Cu_3N 主题内容每年在 Web of Science 发表及被引用统计

统计情况。由图可以看出,近三年来从事 Cu_3N 研究获得的论文成果数量明显增多,科研工作者对 Cu_3N 的关注度快速提升。研究者对 Cu_3N 薄膜如此青睐,归结原因主要来自两个方面:其一,在结构方面,是因为 Cu_3N 晶体组成和结构参数的可调控性,研究者可以通过控制制备技术参数来调配 Cu_3N 晶格常数的大小;或通过在 Cu_3N 晶体中掺入其他不同的原子来改变它的组分,实现 Cu_3N 从绝缘体到半导体甚至导体转变[19-22]。其二,在性能及应用方面,是因为 Cu_3N 的性能及应用方面具有广泛的前景,由于它制备简单、无毒,可取代现有一次性光存储碲基无机相变材料[23-30];由于它较低的热分解温度,有望作为集成电路中金属 Cu 线的缓冲层、低磁阻隧道结的障碍层、自组装材料的模板[28,31-33];由于它具有优异的化学活性,可以应用为新型电池材料和催化剂添加材料[4,7,10,34-36];由于它具有良好的电子发射特性,是场发射材料的有力竞争者[20,27,37,38]。因此,Cu_3N 晶体结构特殊以及物理性能优异,具有诱人的开发及应用前景。

2.1 Cu_3N 薄膜的制备技术

由于铜与氮气都是反应活性小的物质,长期以来,人们认为 Cu_3N 只能用复分解反应制得。1939 年,Juza 等采用高温高压生长技术,用 CuF_2 与氨气反应,制得黑绿色多晶态的氮化铜晶粒[39],尔后有研究者对 Cu—N 系统的电子键态进行了初步的实验研究[40]。约半个世纪后,Zachwieja 等[41]在室温下,利用液氨与 $Cu(NH_3)_4(NO_3)_2$ 混合并加入铜,制得了 1mm 长的 Cu_3N 单晶纳米线。几乎与此同时,Terada 等[42]用磁控溅射单晶外延法制出 Cu_3N 薄膜,接着又在 Pt/MgO 和 Al_2O_3 基底上获得了高定向的外延薄膜。随着 Cu_3N 结构和性能越来越受到关注,以及研究工作的不断深入,很多研究者利用磁控溅射方法,以铜为靶材和氮气作反应气体成功地制备了 Cu_3N 薄膜,该方法也成了制备

Cu_3N 薄膜的主流[1,9,18,26,31,32,37,43-54]。另外,制备 Cu_3N 薄膜的其他方法如分子束外延(MBE)[24,29,42]、反应脉冲激光沉积(ALD)[41,55,56]、射频超声等离子体喷雾法、离子源辅助沉积(IAD)[27,57-59]、磁控溅射离子镀(MSIP)[60]和活化反应蒸发技术(ARE)[61]等也有相关报道,并有研究者利用铜盐或氧化物为原料合成了纯氮化铜纳米晶体[62,63],也有在低温低压下用癸酸癸酯模板作为前体制备了氮化铜纳米颗粒的报道[64]。

制备技术对 Cu_3N 薄膜生长速度的影响较大,制备方法不同,Cu_3N 薄膜的生长速度大相径庭。利用磁控溅射技术制备 Cu_3N 薄膜的生长速率一般在 20nm/min 左右[37,44,58],但也有达到 50~60nm/min 甚至上百 nm/min 的报道[26,43]。其生长速率主要受沉积功率、基片温度、气压(或气压比)、基片与靶的距离、直流或射频方法等的影响,同时也与靶材的形状有很大的关系。有研究者报道,利用磁控溅射薄膜的生长速度不但与气体分压有关,而且它随沉积功率的增加几乎呈线性增大,在功率较小时增大快,而高功率下增长稍慢[44];有人发现在 N_2 分压为 0.13Pa 时薄膜的生长速度最快[37];同时,也有 Cu_3N 生长速率随基片温度的增加而增大的相关报道[45]。

利用磁控溅射技术制备 Cu_3N 薄膜,制备工艺参数包括溅射功率、氮气分压 r、基体温度和沉积气压等主要制备参数对 Cu_3N 薄膜的生长有较大的影响,尤其是氮气分压对 Cu_3N 薄膜生长的晶面取向、晶粒尺寸等的影响很大。通常利用磁控溅射制备的 Cu_3N 薄膜表面致密、光滑、颗粒均匀[35,65-67],典型的 Cu_3N 薄膜表面形貌 SEM 图像,如图 2-2 所示,由于制备工艺条件及参数的不同,Cu_3N 薄膜表面颗粒大小及结晶度有所不同。肖剑荣等[22,44,68,69]利用磁控溅射制备 Cu_3N 薄膜,发现薄膜的形貌特征和生长取向强烈地依赖于气体分压,气体分压增大,薄膜颗粒度明显增大,Cu_3N 由(111)面转向(100)面生长。择优生长取向原因是:低 r 时,主要是由吸附氮原子插入铜原子的晶格形成 Cu—N 键而成膜,薄膜是按照与 Cu(111)晶面一致的晶向生长

Cu₃N(111)面。高氮气分压时,足够的 N 原子与 Cu 在靶表面或基片表面结合成 Cu—N,这时,薄膜按照晶面自由能最低优先生长的原则生长 Cu₃N(100)面。也可解释为:随着氮气分压增大,达到基片表面 Cu 原子数目减少,此时没有足够的 Cu 原子与高能氮结合,导致(100)峰强度减弱[31]。

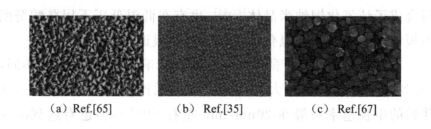

(a) Ref.[65]　　(b) Ref.[35]　　(c) Ref.[67]

图 2-2　Cu₃N 薄膜 SEM 表面形貌图

2.2　Cu₃N 薄膜的晶体结构

20 世纪 70 年代,Terao 首次采用 XRD 技术确定了氮化铜(Cu₃N)具有特殊的反 ReO₃ 晶体结构[70],其空间点群 Pm3m,晶格常数为 0.382nm 左右[43,71,72],但也有最小为 0.3385nm 的结果报道[57],由 TEM 直接测得 Cu₃N 薄膜晶格常数为 0.382nm,如图 2-3 所示。Cu₃N 的晶胞结构如图 2-4 所示,该晶体结构中,Cu 原子位于晶胞的(000)(0,1/2,1/2)(1/2,0,1/2)(1/2,1/2,0)位置,N 原子位于晶胞的(1/2,1/2,1/2)位置,即面心立方结构 Cu(晶格常数 0.3615nm)的八面体间隙位置插入 N(面心立方的八面体间隙是最大间隙位置),可理解为反 ReO₃ 结构的氮化铜体心空位插入 Cu 形成[25]。Cu₃N 每个氮为 8 个晶胞所共有,每个铜原子为 4 个晶胞所共有,分子量为 204.63,块体 Cu₃N 晶体的密度为 5.84g/cm³。实验测得的 Cu₃N 晶体中 Cu—N 键长约为 0.191nm,而 Cu^+ 和 N^{3-} 的半径之和为 0.267nm,而 Cu 原子和 N 原子的半径之和为 0.187nm,因此,Cu 原子和 N 原子之间趋近于以共价键结合,属亚稳共价化合物。也有人认为,Cu₃N 结构中的 Cu 和 N 的结合是以共价键和离子键混合而成[73]。实验制备的 Cu_3N 薄膜常温下呈现为红褐色或墨绿色,其颜色会随着制备技术和工艺参数的不同,导致薄膜中铜和氮原子的化学配比变化,而出现不同的颜色。

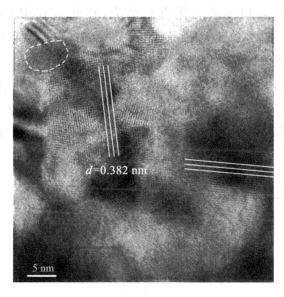

图 2-3 氮化铜晶体的透射电镜图片(Ref.[72])

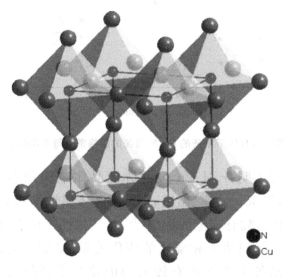

图 2-4 氮化铜晶体反三氧化铼结构示意图(Ref.[23])

制备工艺参数影响 Cu_3N 薄膜的晶格常数,特别是氮气分压对薄膜的结构有至关重要的作用。吴志国等[26]采用柱状靶多弧直流磁控溅射法制备 Cu_3N 薄膜,发现薄膜由立方反 ReO_3 结构的 Cu_3N 和 Cu 的纳米微晶复合而成;随着氮气分压增大,Cu_3N 相逐渐结晶析出,Cu 相逐渐弱化直至消失,在此过程中 Cu 相晶格产生了畸变,晶格常数变小。而肖剑荣等[22]研究结果与此有所不同,Cu_3N 薄膜的晶格常数先随氮气分压增大而增大,当氮气分压增加到一定程度后,Cu_3N 的晶格常数随 r 的增大而略有减小,如图 2-5 所示。Wang 等[37]利用射频磁控溅射在玻璃基片上制备了 Cu_3N 薄膜,同时发现薄膜的晶格常数随着氮气分压的增加而增大。

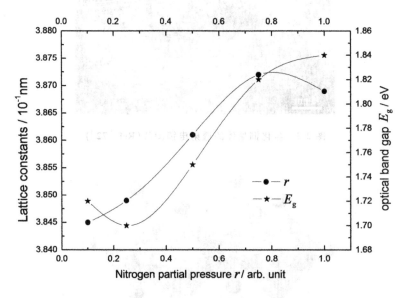

图 2-5　不同氮气分压下制备的 Cu_3N 薄膜的晶格常数与光学带隙(Ref.[22])

众多研究表明,r 对 Cu_3N 薄膜的晶体结构和参数影响具有决定性作用,但沉积功率、温度、气压和基体属性等对 Cu_3N 薄膜的晶体结构及其参数有一定的影响[31,45,74,75]。图 2-6 给出了 Cu_3N 薄膜晶格常数与沉积功率和温度的变化关系,显而易见,Cu_3N 薄膜晶格常数与制备工艺参数有较大的依赖关系,同时,薄膜的电学性能也发生相应的改变。Borsa 等[24]利用 MBE 技术制备了

Cu₃N 薄膜,重点研究了基体和沉积温度对薄膜结构的影响,发现当沉积温度较低时,均可外延 γ′-Fe₄N 和 MgO(100)面结构,但当温度达到 250℃以上时,薄膜为非晶结构。

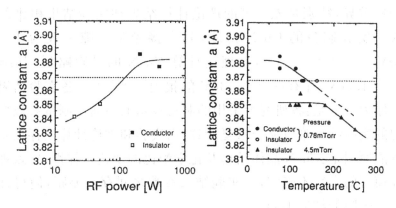

图 2-6 Cu₃N 薄膜的晶格常数随制备功率和温度的变化关系(Ref.[43])

2.3 电学和光学性能

Cu₃N 晶体中,由于铜原子 s 能带与氮原子 p 能带交叠,形成满带,所以纯 Cu₃N 晶体是绝缘体。非常有趣的是,在氮化铜晶体中,铜原子并没有占住其晶格(111)面的紧密位置,而在其晶体结构中留下了许多空隙(结构缺陷),这就容易导致被别的原子填充到 Cu₃N 晶格的中心空位,从而引起 Cu₃N 物理化学性质的显著变化[65,76],这种特殊"结构缺陷"使得 Cu₃N 具有巨大的研究价值,尤其是进行元素掺杂,如碱金属 Li[77,78],过渡金属 Ti[53,69,77,79-81]、Pb[73,82,83]、Ni[80,84]、Zn[80,85]、Cr[48,80,86]、Fe[14,84]、Mn[48,87]、La[88]、Al[52]、Sc[1]和非金属 H[9,89]以及 O[90]等,通过元素掺杂,可以对 Cu₃N 薄膜的物理性能进行有效调控。有研究者采用线性缀加平面波方法(LAPW)计算获得 Cu₃N 掺杂前后的能隙及电子态密度,如图 2-7 所示。显然,掺杂后其电子结构发生变化,费米能明显增加,在费米面 R 线附近电子能带出现重叠,掺杂后 Cu₃N 电学性质发生了改变。

Cu₃N 薄膜的电学性能和光学带隙大小的决定因素及变化趋势几乎一致,主要受其制备工艺参数和元素掺杂的影响。Cu₃N 薄膜的电导率和光学带隙的实验研究结果不同的报道之间还存在一些差异,特别是光学带的理论计算结果和实验结果相比差异较大。实验制备的 Cu₃N 薄膜,其直接光学带隙一般为 $1.1 \sim 1.8 eV$[22,29,37,43,44,46,58,68],而理论计算 Cu₃N 的带隙略低,其结果大致为 1.0eV,并有低至 0.13eV 的报道[7,23,34,73]。各实验结果不同的原因可能是:各研究者采用的制备技术的不同或制备参数变化引起的实验结果难以完全重复所致;而实验和理论计算上存在差异的可能原因是:Cu₃N 晶体结构及化学配比的变化多样性,导致理论计算模型的参数不确定,与实验结果参数之间存在差距,使得计算出光学带隙的结果不同。

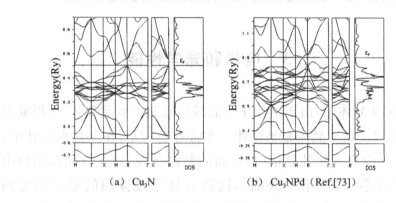

(a) Cu₃N (b) Cu₃NPd (Ref.[73])

图 2-7 LAPW 方法计算能隙和电子态密度

众多研究表明,氮气分压对 Cu₃N 薄膜的电学和光学性能影响很大[18,22,31,49,67,68,71],Cu₃N 薄膜光学带隙与气体分压关系变化曲线,见图 2-5 所示。我们利用射频磁控溅射技术在室温下制备了 Cu₃N 薄膜,研究发现 Cu₃N 薄膜的电阻率随着氮气分压增加而增大,但其在低氮气分压下增加缓慢,在高氮气分压下增长很快[44],其他研究者也报道了类似结果[67,91-93]。产生这种现象的原因:在低氮气分压下,较多被溅射出来的单质铜原子还来不及在靶表面和基底上与氮反应,就沉积在基片上。因此,Cu₃N 薄膜的形成主要靠吸附氮原子插入铜原子的晶格形成 Cu—N 键生长,这时,沉积在薄膜表面有相当一部分的铜原子没有与氮结合成键。所以,Cu_3N 晶格的中心空位仍含有较多的铜原子,这些填隙的 Cu 原子提供了弱局域化电子,而这些弱局域化电子相对以共价键结合的 Cu_3N 晶格上的电子来说,具有非局域性,它的存在能改变膜内电子态密度分布,此时薄膜完全具有半导体甚至导体的特性,它的电阻率小。当氮气分压很高时,高浓度的氮氛围使得有足量的 N 原子与 Cu 原子发生反应形成 Cu—N 键,导致薄膜中不再存在填隙的 Cu 原子,形成电阻率相对较高、甚至接近绝缘的 Cu_3N 薄膜。

2.4 热、力学性能和耐腐蚀性

实验研究表明,Cu_3N 薄膜在室温下十分稳定,在湿度 95%、温度 60℃的情况下放置 15 个月,薄膜的结构和光学性能均未发生变化。但 Cu_3N 薄膜在 100～470℃的温度下很容易分解为 N_2 和 Cu,其分解温度差异很大,是由制备技术和制备工艺参数不同所引起。Cu_3N 薄膜的分解反应方程式如下:

$$2\,Cu_3N \xrightarrow{\Delta} 6Cu + N_2 \uparrow$$

Liu 等[97]在室温下制备的 Cu_3N 薄膜,在真空中经过 100～300℃退火 0.5h,再自然冷却 1.5h,发现膜中只出现微弱 Cu 衍射峰。这些研究报道说明 Cu_3N 薄膜具有较好的热稳定性,但热分解温度较低。

Cu_3N 薄膜的显微硬度没有一个确定的值。Fendrych 首次报道 Cu_3N 薄膜显微硬度和杨氏模量分别达到 8.8GPa 和 146GPa,均随着制备功率的增大而减小[57];有研究表明,氮气分压影响薄膜的硬度,其硬度是基于薄膜中的氮化铜纳米晶体与微小的铜纳米晶粒的复合结果[98]。Chwa 等[99]详细研究了不同温度下制备的 Cu_3N 薄膜与基体 SiO_2 的黏附性,发现在室温和 400℃下沉积的薄膜比在 150℃沉积的薄膜与基体的附着性能差些,但在合适的条件下制备的 Cu_3N 薄膜与基体仍有良好的黏附性[19,47]。

铜抗酸能力很强,但 Cu_3N 薄膜抗酸腐蚀能力极弱,极易溶于稀盐酸溶液。在 $10\sim30g/mol$ 稀盐酸中,Cu_3N 薄膜的腐蚀速率是铜膜的 270 倍,在 $100g/mol$ 稀盐酸中,Cu_3N 薄膜的腐蚀速率是铜膜的 3900 倍;但 Cu_3N 薄膜的耐碱性能力较强,把 Cu_3N 薄膜样品置于 NaOH 碱性溶液中,发现其不溶解。

2.5 Cu_3N 薄膜的应用

1. 光储存及能源材料

Cu_3N 薄膜在较低温度下分解 N_2 和金属 Cu 的特性说明它可以直接进行金属化反应[21,100]。实验研究发现,利用 Cu_3N 薄膜分解后得到铜膜与直接利用磁控溅射得到的铜膜的反射光谱非常相近,并且 Cu_3N 的反射系数比铜的要小,在波长 800nm 左右时反射系数有较大的差别,因此 Cu_3N 薄膜可作为光储存媒介[25,26,61,91]。利用 Ar 离子或激光电子束可以十分方便地在 Cu_3N 薄膜上刻蚀出铜点,如图 2-8 所示。利用 Cu_3N 薄膜这一特性,它又可以被用作光数据存储材料[41,73,77,79,91,101−103]。

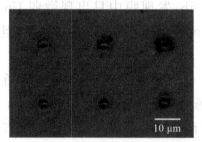 

(a) 电子束写入表面积为3mm×3mm 和1mm×1mm后的显微镜图像,在20kV电子束的曝光区域中观察到点阵列 (Ref.[28])

(b) 激光写入Cu_3N薄膜上的数据,线速度3.5m/s,脉冲功率28mW,脉冲长度250、300、350和400ns,单个的最小直径为0.9mm (Ref.[91])

图 2-8 在 Cu_3N 薄膜上刻蚀出的铜点

太阳能电池是新型半导体材料在应用领域的关键研究目标之一，Cu_3N 薄膜在这一应用领域极具潜力。通过调控制备工艺参数达到调控 Cu_3N 薄膜的晶体结构及薄膜中化学成分配比，从而优化薄膜的光学带隙使得光伏电压达到最大值[12,52,85,88,104,105]。采用现代制备技术在诸如 $SrTiO_3$(100)等基体上沉积绝缘的 p 型和 n 型 Cu_3N(100)薄膜，通过控制制备工艺合理调制薄膜中 Cu/N 的化学配比，实现在 Cu 缺陷条件下进行 p 型掺杂或 n 型掺杂，形成间接带隙符合需求的光伏材料，Cu_3N 薄膜这种双极掺杂特性可提高太阳能转换效率[12,33,103,106]。Cu_3N 薄膜可以作为新型二次电池材料，特别是锂离子电池材料方面具有良好的循环寿命和倍率性能[7,13,36]，这主要是因为 Cu_3N 的电化学包含多个并行过程，除了可逆的锂/氮化铜转化和在纳米复合材料表面的有机层的形成/分解之外，在高循环和高温下促进了可逆的锂/铜氧化物转化过程。

2. 纳米器件、电极及其他材料

随着生活与科技的进步，人们对电子产品的精细度要求越高，纳米器件优势越来越明显。Cu_3N 薄膜可以通过激光照射形成纳米线或点阵，它比普通的纳米棒、纳米线或纳米模板更优越[27,96,101]。Cu_3N 与铁磁导体材料 γ'-Fe_4N 的晶格较匹配，Brosa 等[24]利用外延制备的 Cu_3N/γ'-Fe_4N 双层膜平整光滑，证实该膜可用作集成电路的低磁阻隧道层材料。也有研究者设计制备了基于 Cu_3N 的电阻随机存取存储器(RRAM)，发现其具有良好的电阻切换特性及使用寿命[59]。Cu_3N 薄膜凭借其优异的光电化学性能，显示其具有良好的首次放电量和循环性能，可作为在金属基底的薄膜电极[4,107]。使用掺杂 Cu_3N 纳米颗粒作为芯，原子铂层作为壳，用氮化碳纳米管作为载体，这种催化剂材料可改善氮化物纳米颗粒的形态和分散性，极大提高催化剂性能[10]。可运用磁控溅射技术制备基于 Cu_3N 薄膜的电阻随机存储器件，该器件具有低操作电压和明显差异的电阻率的两级特性等[30,108]。

第3章 氮化铜薄膜的制备

3.1 磁控溅射技术

溅射技术属于PVD(物理气相沉积)技术的一种,是一种重要的薄膜材料制备方法。它是利用带电荷的粒子在电场中加速后具有一定动能的特点,将离子引向欲被溅射的物质制成的靶电极(阴极),并将靶材原子溅射出来使其沿着一定的方向运动到衬底并最终在衬底上沉积成膜的方法。磁控溅射是把磁控原理与普通溅射技术相结合利用磁场的特殊分布控制电场中的电子运动轨迹,以此改进溅射的工艺。磁控溅射技术已经成为沉积耐磨、耐蚀、装饰、光学及其他各种功能薄膜的重要手段。

3.1.1 磁控溅射的工作原理

溅射沉积原理是指用有一定能量的粒子轰击靶材表面,使该靶材表面的原子或者分子离开其表面,发生非常复杂的级联碰撞现象,如图3-1所示。磁控溅射就是以磁场束缚和延长电子的运动路径,改变电子的运动方向,提高工作气体的电离率和有效利用电子的能量。具有低温、高速两大特点。

电子在加速的过程中受到磁场洛伦兹力的作用,被束缚在靠近靶面的等离子体区域内。

电子的运动轨迹将沿电场方向加速,同时绕磁场方向螺

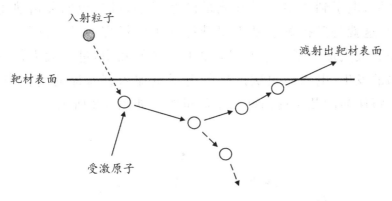

图 3-1 溅射沉积级联碰撞模型

旋前进的复杂曲线。即磁场的存在将延长电子在等离子体中的运动轨迹,提高了它参与原子碰撞和电离过程的概率,因而在同样的电流和气压下可以显著地提高溅射的效率和沉积的速率。

具体说来,磁控溅射系统在真空室充入 0.1～10Pa 压力的惰性气体(Ar),作为气体放电的载体,阴极靶材的下面放置100～1000高斯强力磁铁。在高压作用下 Ar 原子电离成为 Ar^+ 离子和电子,产生等离子辉光放电,电子在加速飞向基片的过程中,受到电场产生的静电作用力和磁场产生的洛伦兹力的共同作用(正交电磁场作用),产生漂移,并做跳栏式的运动。这会使电子到达阳极前的行程大为延长,在运动过程中不断与 Ar 原子发生碰撞,电离出大量的 Ar^+ 离子。磁控溅射时,电子的能量充分用于碰撞电离,使等离子体密度比二极溅射的密度提高约一个数量级。一般靶材刻蚀速率,相应的镀膜速率与靶面电流密度成正比,于是磁控溅射的镀膜速率相比一些普通溅射技术大大提高。经过多次碰撞后电子的能量逐渐降低,摆脱磁力线的束缚,最终落在基片、真空室内壁及靶源阳极上。而 Ar^+ 离子在高压电场加速作用下,与靶材撞击并释放出能量,导致靶材表面的原子吸收 Ar^+ 离子的动能而脱离原晶格束缚,呈中性的靶原子逸出靶材的表面飞向基片,并在基片上沉积形成薄膜。

由于电子必须经过不断地碰撞才能渐渐运动到阳极,而且由于碰撞,电子到达阳极后其能量已经很小,对基板的轰击热也就不大,这就是磁控溅射基板温升低的主要机理。另一方面,加上磁场后大大加大了电子与 Ar 原子碰撞的几率,进而大大促进了电离的发生,电离后再次产生的电子也加入到碰撞的过程中,从而能将碰撞的几率提高好几个数量级。如图 3-2 所示。

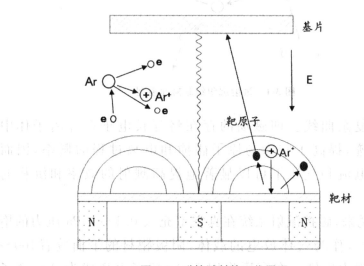

图 3-2　磁控溅射的工作原理

技术上,磁控溅射可以分为直流(DC)磁控溅射、中频(MF)磁控溅射、射频(RF)磁控溅射。三种分类的主要对比见表 3-1。

表 3-1　直流、中频和射频磁控溅射应用对照表

分类	DC	MF	RF
电源价格	便宜	一般	昂贵
靶材	圆靶/矩形靶	平面靶/旋转靶	实验室一般用圆平面靶
靶材材质要求	导体	无限制	无限制
抵御靶中毒能力	弱	强	强
应用	金属	金属/化合物	工业上不采用此法
可靠性	好	较好	较好

3.1.2 磁控溅射主要工艺参数

磁控溅射技术工艺参数较多,通过调节制备工艺参数,可以实现对薄膜结构和性能的调控,影响成膜结构与性能的主要参数如下:

1. 溅射功率

在一定的条件下,溅射功率的增加,会使放电载体如氩气的电离度提高,增加离子的密度,提高溅射速率,并使溅射出来的离子具有较高的能量,从而提高薄膜/基体的附着力及薄膜的致密度。相反,溅射功率太低,离子密度小,沉积速度慢,且离子能量低,得到的薄膜结构疏松,膜层附着力差。

但是并不是溅射功率越大越有利于薄膜沉积。溅射功率过高会使溅射离子动能大大增加,过高的离子能量会产生较大的基底热效应,还会对薄膜造成损伤,薄膜质量下降。这是因为溅射功率较大时,电离得到的离子具有很高的能量,离子打入靶材的深度增加,能量损失增加,使被溅射原子的逸出难度增加,靶材原子不易逸出,使沉积速率降低。同时,由于溅射功率的增加,使溅射时产生的二次电子增多,对基体会产生一定的加热作用,使基体上沉积的 CN 基团挥发,也会降低沉积速率。

2. 基体温度

提高基体温度有利于薄膜和基体间原子的相互扩散,而且会加速化学反应,从而有利于形成扩散结合和化学键附着,使附着力增加。当基体温度较低时,形成膜的原子活性受到限制,形核密度较低,在界面处容易产生孔隙,形成不完全致密的薄膜;而随着基体温度升高,基体表面活性增强,形核密度变大,沉积速率增加,界面孔隙减少,薄膜/基体界面结合较强,附着性变好。

但基体温度过高会使薄膜晶粒粗大,薄膜中热应力增大,薄膜开裂及剥落倾向变大,从而降低薄膜的质量及使用性能。因此要综合考虑基体温度的影响,针对不同的薄膜/基体选择合适的基体温度,得到较好附着性能的薄膜。

3. 溅射气体压强

以常用的 Ar 气为例,Ar 气被电离成 Ar 离子轰击阴极靶材表面,但仍有一部分 Ar 离子混入溅射出的靶原子,沉积到基体表面。因此,如果 Ar 气中杂质过多,膜层中将形成很多缺陷,从而使薄膜结构疏松,降低其表面力学性能,严重影响薄膜质量。

Ar 气分压大小也是影响薄膜质量的重要因素。溅射压力较小时,溅射出来的原子和气体分子的碰撞次数减少,损失的能量较小,可以提高沉积原子与基体的扩散能力,从而提高薄膜的致密度和附着性;如果溅射气体的压力太小,则溅射靶材原子数目较少,薄膜沉积速率降低,且不能起辉或起辉不足。

但如果溅射气压过高,靶材原子与气体的碰撞次数增加,损失能量过多,造成沉积基体的靶材原子能量过低,影响膜层的致密性和附着力。

4. 偏置电压

根据磁控溅射基片即工件偏置电压的不同作用,可分为直流负偏压、脉冲负偏压、交流偏压、零偏压与悬浮偏压五个类别。在基片上加负偏压后,基-阳极间可产生更大的电场力,可使等离子体中的正离子获得更大的能量和加速度轰击基片和工件;可对从靶材表面被溅射出来的原子或分子团等带电粒子进行某种程度的导向和沉积,绕镀性好;在基片和工件上施加不同的负偏压可以消除基片和工件膜层表面在不同的真空度条件下形成的锥状晶和柱状晶;在工件上施加交流偏压,可以中和绝缘膜层上积累的正电荷,减少和消除工件表面打弧;在工件上施加脉冲偏压,因其占空比可连续调节,可以在一定程度上调节工件表面温升。

基片电位直接影响入射的电子流或离子流。基片有目的地选择与施加不同的偏压、选择合适的幅值或占空比、使其按电的极性接收电子或正离子,不仅可以净化基片,增强薄膜附着力,而且还可以改变薄膜的结晶结构。基片选用和施加何种偏置电压对溅射、沉积及镀膜的工艺过程和薄膜质量可以产生严重影响。

如果偏压的类别和参数(电流、电压与占空比)选择合适,膜层的品质和性能可以大为改善。

5. 靶材的影响

靶材作为一种具有特殊用途的材料,具有很强的应用目的和明确的应用背景。脱离开溅射工艺和薄膜性能来单纯地研究靶材本身的性能没有意义。而根据薄膜的性能要求,研究靶材的组成、结构、制备工艺、性能,以及靶材的组成、结构、性能与溅射薄膜性能之间的关系,既有利于获得满足应用需要的薄膜性能,又有利于更好地使用靶材,充分发挥其作用,促进薄膜技术应用的发展。靶材的工艺指标主要包括纯度和结构均匀性。

3.1.3 磁控溅射工艺常见问题

1. 靶中毒

磁控溅射中最常见的问题就是靶中毒。靶中毒时,靶面形成一层绝缘膜,正离子到达阴极靶面时由于绝缘层的阻挡,不能直接进入阴极靶面而是堆积在靶面上,容易产生冷场致弧光放电——打弧,使阴极溅射无法进行下去。

靶中毒的因素主要是反应气体和溅射气体的比例,反应气体过量就会导致靶中毒。反应溅射工艺进行过程中靶表面溅射沟道区域内出现被反应生成物覆盖或反应生成物被剥离而重新暴露金属表面此消彼长的过程。如果化合物的生成速率大于化合物被剥离的速率,化合物覆盖面积增加。在一定功率的情况下,参与化合物生成的反应气体量增加,化合物生成率增加。如果反应气体量增加过度,化合物覆盖面积增加,如果不能及时调整反应气体流量,化合物覆盖面积增加的速率得不到抑制,溅射沟道将进一步被化合物覆盖,当溅射靶被化合物全部覆盖的时候,靶完全中毒。

靶中毒的物理机制:一般情况下,金属化合物的二次电子发射系数比金属的高,靶中毒后,靶材表面都是金属化合物,在受到离子轰击之后,释放的二次电子数量增加,提高了空间的导通能力,降低了等离子体阻抗,导致溅射电压降低,从而降低了溅射速率。一般情况下磁控溅射的溅射电压为 400～600V,当发生靶中毒时,溅射电压会显著降低。金属靶材与化合物靶材本来溅射速率就不一样,一般情况下金属的溅射系数要比化合物的溅射系数高,所以靶中毒后溅射速率低。反应溅射气体的溅射效率本来就比惰性气体的溅射效率低,所以反应气体比例增加后,综合溅射速率降低。

靶中毒的解决办法:采用中频电源或射频电源;采用闭环控制反应气体的通入量;采用孪生靶;控制镀膜模式的变换:在镀膜前,采集靶中毒的迟滞效应曲线,使进气流量控制在产生靶中毒的前沿,确保工艺过程始终处于沉积速率陡降前的模式。

2. 膜质量问题

磁控溅射镀膜时,由于操作或处理不当会导致膜质量下降,其常见问题及主要解决方式如下:

(1)膜层灰暗及发黑:

真空度低于 0.67Pa,应将真空度提高到 0.13～0.4Pa。

氩气纯度低于 99.9%,应换用纯度为 99.99% 的氩气。

充气系统漏气,应检查充气系统,排除漏气现象。

底漆未充分固化,应适当延长底漆的固化时间。

镀件放气量太大,应进行干燥和封孔处理。

(2)膜层表面光泽暗淡:

底漆固化不良或变质,应适当延长底漆的固化时间或更换底漆。

溅射时间太长,应适当缩短。

溅射成膜速度太快,应适当降低溅射电流或电压。

(3)膜层色泽不均:

底漆喷涂得不均匀,应改进底漆的施涂方法。

膜层太薄,应适当提高溅射速度或延长溅射时间。

夹具设计不合理,应改进夹具设计。

镀件的几何形状太复杂。应适当提高镀件的旋转速度。

(4)膜层发皱、龟裂：

底漆喷涂得太厚，应控制在 7～10mm 厚度范围内。

涂料的黏度太高，应适当降低。

蒸发速度太快，应适当减慢。

膜层太厚，应适当缩短溅射时间。

镀件温度太高，应适当缩短对镀件的加温时间。

(5)膜层表面有水迹、指纹及灰粒：

镀件清洗后未充分干燥，应加强镀前处理。

镀件表面溅上水珠或唾液，应加强文明生产，操作者应戴口罩。

涂底漆后手接触过镀件，表面留下指纹。应严禁用手接触镀件表面。

涂料中有颗粒物，应过滤涂料或更换涂料。

静电除尘失效或喷涂和固化环境中有颗粒灰尘，应更换除尘器，并保持工作环境的清洁。

(6)膜层附着力不良：

镀件除油脱脂不彻底，应加强镀前处理。

真空室内不清洁，应清洗真空室。值得注意的是，在装靶和拆靶的过程中，严禁用手或不干净的物体与磁控源接触，以保证磁控源具有较高的清洁度，这是提高膜层结合力的重要措施之一。

夹具不清洁，应清洗夹具。

底涂料选用不当，应更换涂料。

溅射工艺条件控制不当，应改进溅射镀工艺条件。

3.2 JGP-450a 型多功能磁控溅射系统

氮化铜的制备方法较多，应用较多且比较成熟的方法是利用射频磁控溅射的方法，本书采用中国科学院沈阳科学仪器研制中心有限公司制造的 JGP-450a 型多功能磁控溅射设备(三靶位)制备氮化铜薄膜样品。

3.2.1　JGP-450a 型设备的主要组件

JGP-450a 型多功能磁控溅射设备为双套不锈钢真空室结构，配置 600L/S 分子泵机组一套，微机型复合真空计一台，质量流量控制显示器一台，5 只 2 英寸的永磁磁控靶、1 只 3 英寸可镀磁性材料的专用磁控溅射靶，真空室配有可加热衬底从室温到 600℃ 的自旋转带挡板样品台两套，烘烤照明系统两套。可以采用单靶独立、双靶或三靶任意轮流或组合(需配 2～3 个溅射电源)共溅工作模式，射频直流兼容。溅射方向采用由下向上，向心溅射方式，这样可以避免微粒物质落到基片上进而提高镀膜质量。适用于各种单层膜、多层膜及掺杂膜的镀制。

JGP-450a 型设备主体均为优质不锈钢制造，耐腐蚀、抗污染、漏率小；设备电控部分采用了先进的检测和控制系统，量值准确，性能稳定、可靠；设备布局合理，提高了射频源的利用率和稳定性，减少了对环境的射频干扰；控制面板的设计考虑了美观和适用的结合，使面板操作指示明确，观察舒适、操作方便。设备的基片加热温度、靶头与基片的距离、充入气体的流量、基片架的旋转速度、射频电源的输出功率均实现无级调整；基片与靶头的定位精度可达 1mm。

基片加热采用进口金属铠装丝加热，加热速度迅速、均匀、热效率高，而且对真空室无污染。

溅射用的靶材可以是导电材料也可以是绝缘陶瓷材料。

磁控溅射靶面在超高真空的环境中，沿轴向与基片距离在线动态连续可调，这一功能为寻找最佳实验参数提供了强有力的实验手段。为了操作的方便性，每只靶上都配有电机，可在操作面板上电动控制靶面到基片的距离。

样品台与靶之间可进行双向调整。

JGP-450a 型设备的主要组件及技术指标如下：

极限真空度：$\leqslant 6.67 \times 10^{-5}$ Pa（经烘烤除气）；$\leqslant 6.67 \times 10^{-6}$ Pa（进口分子泵）；

系统真空检漏漏率：$\leqslant 5.0 \times 10^{-7}$ Pa.L/S；

系统短时间暴露大气并充干燥氮气后，再开始抽气，40min 真空度可达到 6.6×10^{-4} Pa；

停泵关机 12h 后真空度：$\leqslant 5$ Pa；

系统主要由溅射真空室、磁控溅射靶、基片水冷加热台、工作气路、抽气系统、安装机台、真空测量及电控系统等部分组成。

(1) 溅射真空室组件。

圆筒形真空室尺寸为 $\phi 450mm \times 350mm$，电动上掀盖结构，可内烘烤 $100 \sim 150°C$，选用优质不锈钢材料制造，氩弧焊接，表面进行特殊工艺抛光处理，接口采用金属垫圈密封或氟橡胶圈密封；真空室组件上焊有各种规格的法兰接口 15 个。

(2) 磁控溅射靶组件。

靶材尺寸：$\phi 60mm$（其中一个可以溅射磁性材料）；

永磁靶：射频溅射与直流溅射兼容，靶内有水冷；

电动控制挡板组件：1 套；

靶配有屏蔽罩，以避免靶材之间的交叉污染（非共溅时安装使用）；

三个靶可共同折向上面的样品中心，靶与样品距离 $90 \sim 110mm$ 可调；当直接向上溅射时，靶与样品距离 $40 \sim 80mm$ 可调，并有调位距离指示。

(3) 单基片水冷加热台组件

基片尺寸：可放置 $\phi 30mm$ 基片；

基片加热最高温度 $600°C \pm 1°C$，由热电偶闭环反馈控制；

取下加热炉可以换上水冷基片台；

基片可连续回转，转速 $5 \sim 10 r/min$；

基片可以加负偏压 $-200V$；

手动控制样品挡板组件：1 套；

样品托:2个。

(4) 抽气机组及工作气路

复合分子泵及变频控制电源:1台(抽速1400L/s);

直联机械泵:1台(抽速8L/s);

200SCCM、100SCCM质量流量控制器、充气阀RF16、进气截止阀、混气室、管路、接头等:共2路;

充气阀D6,管路、接头等:1路(解除真空充氮气)。

(5) 真空测量及电控系统

电源机柜:2台(含供电电源,有断电保护功能);

控制电源:1台(为机械泵、电磁阀、升降电机等提供电源);

水流报警系统:1套(对分子泵、磁控靶有断水报警切断相应电源功能);

样品加热控温电源:1套(可实现程序控温);

电机控制电源:1套;

靶挡板电源:3套;

加热烘烤及照明电源:1套;

宽量程数显真空计:1套(测量范围:$1.0 \times 10^5 \sim 1.0 \times 10^{-5}$ Pa。)

200SCCM、100SCCM质量流量显示器:1套;

分子泵控制电源:1套;

500W射频电源:1套;

500W直流电源:2套;

-200V直流偏压电源:1套。

(6) 计算机控制系统

控制的内容主要有靶挡板、样品自转、六工位样品挡板、样品控温等。

3.2.2　JGP-450a型设备的结构

JGP-450a型磁控溅射系统是中国科学院沈阳科学仪器研制中心有限公司设计并制造,桂林理工大学根据科研工作的需要在真空室内部和气路上进行了部分改造,其外观如图 3-3 所示。

图 3-3　JGP-450a 型多功能磁控溅射设备外观图

JGP-450a 型磁控溅射系统主机结构简图如图 3-4、图 3-5 所示。图 3-4 为正面图,图 3-5 为俯视图。

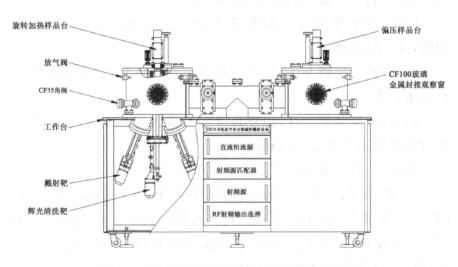

图 3-4　JGP-450a 型多功能磁控溅射设备结构正面图

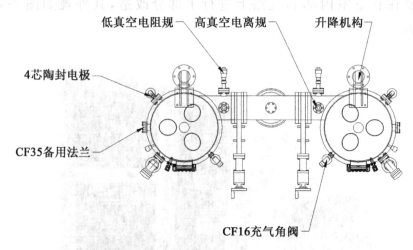

图 3-5　JGP-450a 型多功能磁控溅射设备结构俯视图

3.2.3　设备操作流程

（1）操作设备前，首先检查各种阀门是否全部处于关闭状态。如不是关闭状态，需重新置于关闭状态。

（2）打开水源，确定各路水路是否畅通、有无渗漏。如发现有问题，要及时解决。

（3）打开总电源开关，检查三相指示是否正常，其他电源应都在关闭状态。

（4）打开复合真空计，检查真空室内是否有真空度，根据真空度的情况分别采取以下两种抽气方式。

方式一：真空室真空度不低于 20Pa。

此种情况的操作方式为：启动机械泵，启动预抽阀，快速打开 CF 35 旁抽角阀（要全打开），待真空度抽至低于 20Pa 时，先关闭 CF 35 旁抽角阀，关闭预抽阀，启动前级阀再打开插板阀（一定要开到位），启动分子泵电源，抽至所需真空度。

方式二：真空室内真空度低于 20Pa。

此种情况的操作方式为：启动机械泵，启动前级阀，打开插板阀(一定要开到位,如遇开启费力,则应立即通知相关人员检查或修理,千万不可用蛮力开启),启动分子泵电源,抽至所需真空度。

(5)真空室抽至所需本底工作真空后,一般情况下,真空度应低于5×10^{-4}Pa,此时可缓慢打开真空室的CF 16充气角阀,待真空度稳定后,对需要使用的电源进行预热。打开所需气体的气瓶及进气电磁阀,打开流量控制显示仪,将选择开关置于阀控档,缓慢调节进气流量,配合插板阀控制抽速,将真空度控制在工艺要求的范围内,此时就可以进行正常的溅射镀膜了。

(6)关闭设备时,要先关闭溅射电源、样品架加热电源,样品架旋转电源,再关闭气瓶,进气电磁阀及流量显示仪,最后关闭CF 16充气角阀,打开插板阀,将真空度恢复到5×10^{-4}Pa时,关闭插板阀。如要取样片,可在确认真空室内温度不高于100℃时打开放气阀,最好能通过放气阀充入干燥氮气。待真空室内与外界气压平衡时,关闭放气阀,启动升降机构。取出被镀样片,最好能同时装上新样片,启动升降机构,落下真空室上盖,再按(4)中方式一操作,将真空度抽至本底真空,关闭CF 150插板阀,关闭分子泵电源,关闭前级电磁阀,待分子泵示数为0时,关停机械泵,关闭总电源开关。15min后,关闭水源。

(7)另外如果需要做连续镀膜实验,就需要在分子泵不停止工作的情况下,开真空室盖,取出被镀样片同时装上新样片。此种情况可采用如下操作方式:关闭溅射电源,样品架加热电源,样品架旋转电源,流量仪,CF 16充气阀,在关闭插板阀后,通过放气阀充入干燥氮气,观察真空计,确保真空度达到一个标准大气压,即1×10^5Pa。打开盖取出样片,装入新样片,盖好真空室盖。接着按照方式一进行操作。

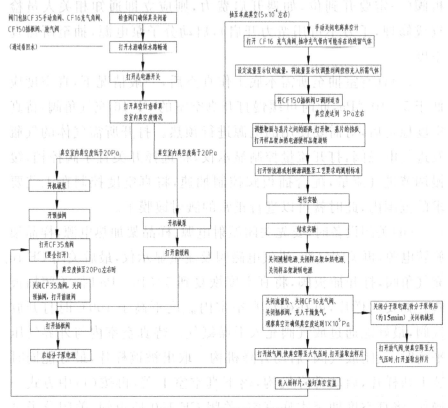

图 3-6 设备操作流程图

注：取出被镀样片时，最好能同时装上新样片，启动升降机构，落下真空室上盖。再按上面介绍的操作方式将真空度抽至本底真空。关闭 CF150 插板阀，关闭分子泵电源，待分子泵示数为 0 时，关停机械泵，关闭总电源开关，15min 后，关闭水源，停止工作。

3.3 氮化铜薄膜的制备

磁控溅射制备氮化铜薄膜，靶材均选用高纯铜靶（99.99%），反应气体是氮气（99.999%），工作气体采用 Ar（99.99%）气体。在掺杂实验中，本书主要研究了 Ti、Ag 等金属掺杂。

3.3.1 实验参数选取

实验参数选取灵活性比较大,薄膜的组分和结构都可以通过控制沉积参数来加以改变。但其缺点是过程有太多的可调参量,这些参量使沉积过程控制复杂化,使实验重复的可能性变低。由于可调参量多,因此在实验时就要合理选择参量,减少实验过程中材料浪费、仪器的损耗,同时节约研究时间。实验中主要选取射频功率、氮气分压 r 和沉积温度为主要研究参量,把氮气分压作为研究重点,沉积时气体压强一般在 0.1~1.0 Pa。各参量的主要取值见表 3-2。在研究过程中,再确定各参量重点研究值(表中黑体数据),如射频功率 50W、100W,沉积温度室温、80℃,氮气分压 r 为 0.25Pa 等,进行交错实验,以求在最少的实验次数下达到最佳的实验效果。譬如,采用全部实验参量实验时,按表中所列数据需要进行的实验次数为: $n=C_6^1 \times C_6^1 \times C_6^1 = 216$ 次,而选定重点研究参量后,重点做的实验次数为: $n=C_3^1 \times C_5^1 \times C_3^1 = 45$ 次,重点实验可能还需重复做。当然,重点研究参量是在多次实验和广泛参考前人研究的基础上得到的。

表 3-2 实验研究的主要研究参量及其取值

研究参量	取值范围	各研究参量主要取值					
射频功率/W	20~300	**20**	50	80	**100**	150	200
氮气分压/Pa	0~1.0	**0.1**	**0.25**	**0.5**	**0.75**	**1.0**	0
沉积温度/(℃)	室温~250	**室温**	**80**	100	**120**	160	200

3.3.2 薄膜的溅射生长

1. 基片处理

实验采用的基片有玻璃片、石英片和单晶 P 型(111)硅片(B 掺杂,电阻率 $\rho=10\sim20\Omega \cdot cm$)等。在沉积前,基片必须加以清洗,微量的玷污都会对样品引起很大的危害,并可以完全改变薄膜的特性:使薄膜的黏附性下降、电学性能变差、内应力变大、在光学上表面呈现出斑驳的油膜。基片清洗工艺的基本准则是消除表面的有机物、颗粒、过渡金属和碱性离子,各种清洗方法在文献[72]上有详细的介绍。本书清洗基片的方法如下:

首先,基片浸泡在丙酮(分析纯)溶液中用超声波清洗机清洗10～15min,再用去离子水冲洗,除去表面的无机物及其他杂质。

然后,将基片浸泡于酒精(化学纯)中,用超声波清洗机清洗10～15min,再用去离子水冲洗,以除去其表面有机污垢。

最后,基片用烘箱烘干,不能让基片自行晾干,否则,在基片上出现水印,尤其对光学性质会产生很大的影响。

沉积前,再在50～100W 的射频功率、4.0～8.0Pa 的气压下,用 Ar 轰击基片表面 10min,以达到更进一步清洗基片表面的作用。

2. 薄膜调控生长

薄膜沉积阶段是整个实验过程的关键阶段,选取好沉积参量,控制其变化范围,可以很好地调制薄膜的组分和结构,获得理想的薄膜。如果样品用来测试 X 射线衍射,一般沉积气压选定略低,在 1.0Pa 以下;如果考虑较大的沉积速度,气压一般调到较高;如果希望薄膜结构紧密且稳定,可适当给基体加热。对于 Cu_3N 薄膜而言,气体分压对薄膜的晶格取向生长影响较大,可根据需要调节气体分压;同时,通过调节气体分压,从而改变薄膜内的组分和键结构,达到改变薄膜的光学、电学等物理性能。沉积时间的选取也是根据需要来决定,沉积时间可根据需要在 10～30min 内变化,如果样品是用来研究光学性能,时间一般为 5～20min,如果用来进行电学性能测试,时间一般为 20～30min。

第4章 氮化铜薄膜的结构研究

材料性能决定于材料的组分与结构。材料科学与工程的任务是研究材料的结构、物理化学性能、加工和使用性能四者之间的关系。对于一定组成的多晶材料来说,就是决定于材料的结构,包括晶体结构和显微结构,也就是说,包含用肉眼和低倍放大镜观测到的材料的宏观组织结构,用光学或电子显微镜观测到的材料的显微组织,用场显微镜观测到的材料的原子象及原子的电子结构等。晶体结构对性能的影响主要体现在两个方面:

(1)材料内部结合键的强度:决定了材料的弹性模量和强度。

(2)材料的各向异性:产生内应力——多数情况下对材料的性能不利。

对于具体的材料来说,材料的显微结构在很大程度上决定材料的强度。事实上,材料的显微结构因素都会对材料性能等产生影响,如形状、晶粒大小、取向,气孔的形状、大小、含量和分布,晶界、晶相、杂质,缺陷(内部、表面、裂纹)等。研究材料的结构,对理解材料的性能具有重大的意义。

4.1 薄膜的结构分析

4.1.1 氮气分压对薄膜结构的影响

图 4-1 所示为不同氮气分压下(沉积功率为 200W)制备薄膜样品的 X 射线衍射图谱。由图可见,随着氮气分压的增大,Cu_3N

薄膜 23.32°处的(100)晶面衍射峰逐渐加强,而 41.05°处的(111)晶面衍射峰逐渐减弱,这说明薄膜在不同的氮气分压下择优生长;在低的氮气分压下 Cu_3N 薄膜择优(111)晶面生长,而在较高的氮气分压下其择优生长沿(100)晶面。在图 4-1(b)和(c)中出现的 Cu_3N 薄膜微弱的(110)衍射峰,而在图 4-1(a)和(d)中几乎没有(110)衍射峰,说明只有在适当的氮气分压 r 下(110)晶面才会生长。这说明氮气分压影响 Cu_3N 薄膜各晶面择优生长取向的规律。原因十分简单:在低氮气分压下,主要是由吸附氮原子插入铜原子的晶格形成 Cu—N 键而成膜,薄膜是按照与 Cu(111)晶面一致的晶向生长 Cu_3N(111)面。在高氮气分压下,有足够的 N 原子与 Cu 在靶表面或基片表面结合成 Cu—N,这时,薄膜按照晶面自由能最低优先生长的原则生长 Cu_3N(100)面。由不同氮气分压下 Cu_3N 薄膜的 XRD 谱,通过仪器计算得到了薄膜的晶格常数,见表 4-1。可见,实验所得的 Cu_3N 薄膜的晶格常数最小值为 3.809Å,最大值仅为 3.867Å,随着氮气分压的减小而增大。

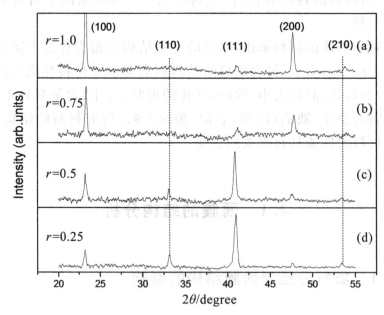

图 4-1 不同氮气分压下 Cu_3N 薄膜的 XRD 图谱

同时发现,它们比其他研究者报道的实验值(3.868Å 左右)偏小[21],这可能是由于样品的制备方法和测试仪器的不同所引起。在 Maruyama 等的报道中[21],Cu_3N 薄膜的晶格常数在 3.868Å 以上时,它为导体,而在 3.868Å 以下时则为导体,这一点与本书中得到 Cu_3N 薄膜电阻率的结果吻合得很好。

表 4-1 不同氮气分压下制备 Cu_3N 薄膜的晶格常数

氮气分压	1.0	0.75	0.50	0.25	0.10
晶格常数/Å	3.809	3.834	3.851	3.864	3.871

4.1.2 溅射功率对薄膜结构的影响

图 4-2 所示是氮气分压为 0.75 时,沉积功率分别为 100W、200W 和 300W 条件下制备的 Cu_3N 薄膜 XRD 谱。从图中可以看出,在 XRD 衍射谱中 23.32°处的(100)和 47.74°处(200)衍射峰随着沉积功率的增加迅速增加,而其他(110)和(111)衍射则几乎没有变化。可见,沉积功率只影响 Cu_3N 薄膜内的晶化程度,而对膜内晶体的生长取向没有影响。

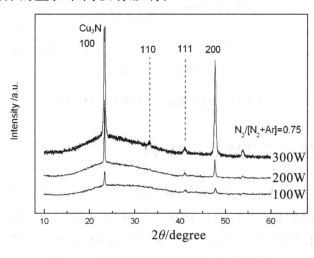

图 4-2 不同沉积功率下薄膜的 XRD 图谱

4.1.3 基片温度对薄膜结构的影响

图 4-3 所示是氮气分压为 0.5 时,沉积功率为 100W,沉积稳定分别为 50℃、150℃ 和 250℃ 条件下制备的 Cu_3N 薄膜 XRD 谱。从图中可以看出,沉积温度对薄膜的生长取向有一定的影响,但是与氮气分压对薄膜生长取向影响相比,沉积温度对薄膜生长取向的影响要小得多。在较低的沉积温度下其择优生长沿(111)晶面,在较高的沉积温度下其择优生长沿(100)晶面。我们认为这种(100)晶面强度随着沉积温度增长的原因是:由于在较高的沉积温度下氮原子在沉积表面获得了较大的能量与铜原子结合,导致了较高浓度的 Cu—N 键的形成。

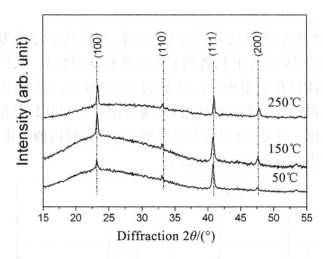

图 4-3 不同沉积温度下薄膜的 XRD 图谱

4.1.4 掺杂后薄膜的 XRD 谱

掺 Ti 前后 Cu_3N 薄膜的 XRD 谱如图 4-4 所示。从图 4-4 可见,Cu_3N 薄膜掺 Ti 后仍具有反三氧化铼结构。在该沉积条件下图谱中没有出现 Cu 峰,说明在薄膜中没有未成键的单质铜。未掺杂的 Cu_3N 薄膜优先生长(111)面,而当掺 Ti 后,薄膜优先生长

(100)面[99]。另外,Cu_3N(100)峰强随着薄膜内 Ti 含量的增加而增强,在图谱中 36.7°和 42.4°峰分别对应于 TiN(111)和 TiN(200)。由于铜-氮之间与钛-氮之间,前者更容易发生反应,这样即使在反应室中只存在比较少的钛,但还是有较多的 Ti—N 键形成。这也说明,在制备的样品中,没有单质的钛元素,所有钛均与氮原子成键。

同样,形成薄膜晶体的晶粒大小可以利用 XRD 进行初步的估算,在这里应用 Debye-Scherre 公式[109]:

$$D_0 = \frac{k\lambda}{\beta_0 \cos\theta} \quad (4-1)$$

式中,D_0 为晶粒大小;k 为常数;λ 为 X 射线波长;β_0 为相应峰的半波宽度;θ 为衍射角度。在图示中估算出的晶粒大小大约为 58nm,比从 AFM 图示中获得的粒子颗粒的大小要偏小。

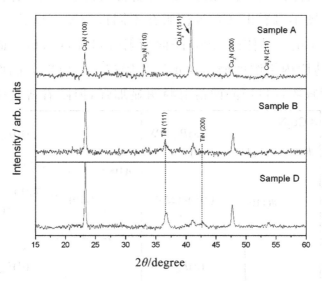

图 4-4 掺 Ti 前后薄膜的 XRD 图谱(样品 A 未掺杂,样品 B、D 掺杂)

利用双靶共溅射技术,制备银掺杂氮化铜薄膜(Cu_3NAg)。其他制备条件不变,通过改变银靶溅射功率 P_w,从而达到改变薄膜中 Ag 的浓度,溅射条件见表 4-2(掺银时,参数条件与该表相同)。

表 4-2　Cu_3NAg 薄膜制备参数

Parameters	Value
Cu target RF power/W	200
Ag target DC power P_w/W	0,3,6,9
N_2 gas flow(SCCM)	40
Ar gas flow(SCCM)	10
Working pressure/Pa	1.0
Deposition temperature/℃	RT
Deposition time/min	15

图 4-5 所示是 Cu_3N 和银掺杂氮化铜薄膜的样品的 XRD 图谱。图中强度最大的几个峰(100)、(111)、(200) 与 Cu_3N(PDF 47-1088)相符合,杂质峰较少,说明制备的薄膜质量较高,相组成单一。由图不难看出,Cu_3N 薄膜在不同银浓度下择优生长:在未掺 Ag 前,Cu_3N 晶体的(100)晶面衍射峰最强,(111)晶面峰位较弱;随着薄膜内银浓度增加,Cu_3N 薄膜(100)和(200)晶面衍射峰强度逐渐减弱,(111)晶面衍射峰明显加强,但随着薄膜中银浓度

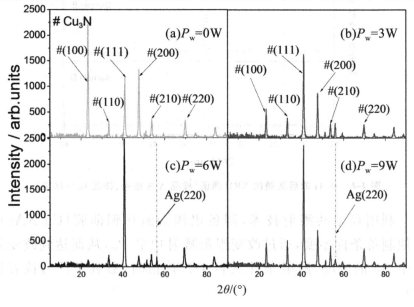

图 4-5　掺 Ag 前后 Cu_3N 薄膜 XRD 能谱图

的进一步增加,(111)又相对减弱;这与其他研究者报道的掺杂或制备条件变化薄膜择优生长相类似[35,48,69,93]。同时,在56°附近的衍射峰为金属 Ag 的(220)晶面衍射产生,证明有部分 Ag 在薄膜之中以金属相形式存在[19]。

4.2 薄膜的表面形貌

制备出的薄膜中由于含有 Cu_3N 和 Cu 两相,因此,薄膜的颜色与这两相密切相关。由于使用的靶材是纯铜,随着 Cu_3N 含量的增多,薄膜能够从 Cu 相的黄色一直变化到 Cu_3N 相的棕黑色;其次,薄膜的颜色深度还与薄膜的厚度有一定的关系。

图 4-6 所示是在沉积功率 200W、氮气分压为 0.25、沉积温度为 80℃条件下沉积的 Cu_3N 薄膜 $3.5\mu m \times 3.5\mu m$ AFM 二维表面形貌图。可见,薄膜表面平紧、致密。薄膜的最高峰值为 27.018nm,粗糙度为 2.545nm,均方根粗糙度为 3.188nm。

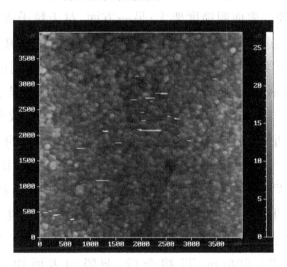

图 4-6 Cu_3N 薄膜 AFM 二维形貌图

图 4-7 所示为不同氮气分压 r 下制备的氮化铜薄膜样品的 AFM 二维表面形貌图。可见,在不同氮气分压下制备的薄膜都比

较致密、均匀,但形貌特征与氮气分压的变化有着强烈的依赖关系,氮气分压升高,薄膜颗粒度明显增大。当氮气分压为 0.1 时,薄膜颗粒大小约为 50nm。当氮气分压为 0.75 时,薄膜颗粒大小约 100nm。在所观察范围内,薄膜的均方粗糙度由 1.181 升高到 1.373。

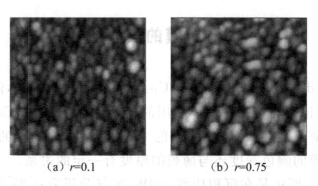

(a) $r=0.1$ (b) $r=0.75$

图 4-7 不同氮气分压下制备的 Cu_3N 薄膜表面 AFM 形貌

薄膜表面形貌变化的原因:一方面,由于氮气分压的增加,薄膜中 Cu—N 相对含量增加,而 Cu 相减少,由于 Cu_3N 的晶格常数比 Cu 大,因此,成膜的表面粗糙度要大;另一方面,对于粒子或原子在薄膜表面生长成膜或刻蚀的双重机制来说,刻蚀作用在低氮气分压下比高氮气分压下要强,因为在较低的氮气分压下,具有刻蚀作用的 Ar 浓度高,较强刻蚀作用可以除去薄膜中部分弱结合,使膜结合紧密,导致颗粒变细,与此同时膜中空隙相对减少,使得膜更加致密均匀。

图 4-8 所示为掺 Ti 前后的氮化铜薄膜样品的 AFM 三维 3000nm×3000nm 表面形貌图(沉积条件相同)。从图中可以看出,掺 Ti 前后薄膜的表面形貌发生了较大的变化:掺杂前薄膜具有金字塔形貌,Ti 掺杂后薄膜具有球形形貌。掺杂前,纯氮化铜薄膜的表面均方根粗糙度(RMS)和最高峰值 R_{max} 分别为 3.188nm 和 23.606nm;Ti 掺杂后,薄膜的表面均方根粗糙度(RMS)和最高峰值 R_{max} 分别为 8.413nm 和 56.840nm。纯氮化铜薄膜的平均颗粒大小在 70nm 左右,而掺 Ti 后薄膜的颗粒大小大约在 100nm。掺 Ti 前后薄膜表面形貌的变化,我们认为可能是

以下两个原因所致：第一，TiN 是面心立方结构，其晶格常数为 0.4242nm[110]，但是 Cu_3N 是反三氧化铼结构，其晶格常数为 0.3815nm[29]；第二，在薄膜的生长过程中，由于 Ti 的掺入改变了薄膜的优先生长取向，纯氮化铜薄膜优先生长(111)方向，而掺杂后薄膜优先生长(100)方向。

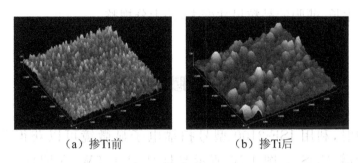

(a) 掺Ti前　　　　　(b) 掺Ti后

图 4-8　掺 Ti 前后薄膜的 AFM 形貌图

图 4-9 (a)(b)(c)给出了 Cu_3NAg 样品扫描电子显微镜图像。从图中可见，未掺杂 Cu_3N 晶体晶粒十分均匀，平整致密、表面粗糙小。比较不同 P_w 下制备 Cu_3NAg 样品的图像发现，随着 P_w 增加，样品晶粒尺寸变大，薄膜表面逐渐变得粗糙。当 P_w 为 6W 时，样品呈现较大

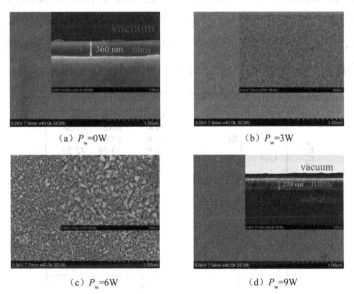

(a) P_w=0W　　　　　(b) P_w=3W

(c) P_w=6W　　　　　(d) P_w=9W

图 4-9　Cu_3N 和 Cu_3NAg 薄膜的 SEM 表面形貌及其截面图

规整的晶体颗粒,表面显得粗糙,结晶度高。当 P_w 继续增大,薄膜表面颗粒又变得细小,表面粗糙度减小。分析原因可能是:Cu_3N 薄膜在成核、长大和成膜的过程中,当 P_w 较小时,掺杂 Ag 随机分布在 Cu_3N 薄膜之中,含量较少,对薄膜的表面形貌影响不大;当 P_w 为 6W 时,掺入薄膜中 Ag 的比例增加到一个合适的浓度,很大程度上影响 Cu_3N 薄膜的生长,薄膜的晶粒尺寸变大,并十分规整。

4.3 薄膜的组分

同时,利用 JSF-2100 型号扫描电子显微镜可以获得薄膜的表面能谱(EDS)。图 4-10 所示是样品 $P_w=9W$ 的 EDS 谱,谱中各元素的原子和质量百分比见图中插图。同时,根据 EDS 测试结果可得到不同 P_w 下制备的 Cu_3NAg 薄膜中各元素原子及质量百分比,见表 4-3。由表可见,首先,随着 P_w 的增大,薄膜中 Ag 原子的原子百分比增加,而铜、碳原子的原子百分比先减少后稍有回升;其次,薄膜表面含有较多的 O,它主要来自薄膜吸附大气中的水分,薄膜中较高的 Si 含量则来自单晶硅基片所致。同时,由表也可发现,当 P_w 在 9W 时,薄膜中 Ag 原子的原子百分比迅速增加,也说明了随着 Ag 溅射功率增加,薄膜内 Ag 的含量迅速增加。

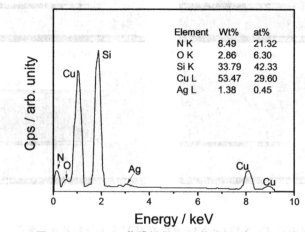

图 4-10　Cu_3NAg 薄膜样品 EDS 能谱图($P_w=9W$)

表 4-3 Cu_3N 薄膜和 Cu_3NAg 薄膜的 EDS 谱数据,原子百分比含量(E) at%

原子	0	3W	6W	9W
Cu	34.28	25.97	24.70	29.60
N	23.66	18.58	18.70	21.32
O	12.72	6.63	9.17	6.30
Si	29.34	48.81	33.62	42.33
Ag	0	0.02	0.15	0.45

图 4-11 所示为不同掺杂功率 P_w 下制备的 Cu_3N 和 Cu_3NAg 薄膜的 XPS 能谱图,由谱可得到膜中各元素的组分与 EDS 结果完全一致。由图 4-11 中插图明显可以看出,在 368.08eV 和 374.08eV 位置有两个明显紧邻峰,它们分别对应于 $Ag3d_{5/2}$ 和 $Ag3d_{3/2}$[111,112]。位于 368.08eV 的峰表明 Ag 以金属相(Ag^0)的形式分布在反三氧化铼结构 Cu_3N 网络中,并以 Ag 纳米团簇的形式存在[113]。鉴于银以单质银(Ag^0)和银离子(Ag^+)的形式存在于 Cu_3N 薄膜中,可以认为它在 Cu_3N 薄膜中可能会担任三个角色:可以在 Cu_3N 结构起到填隙的作用,也可以替代 Cu_3N 晶格中的 Cu,可与膜中的氧结合形成 Ag_2O[114]。

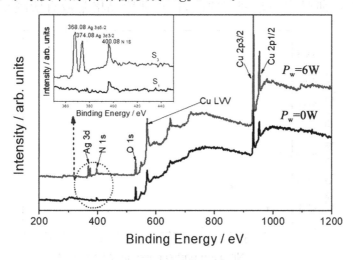

图 4-11 Cu_3N 和 Cu_3NAg 薄膜 XPS 能谱图,插图为 Ag3d 峰

4.4 薄膜的晶格常数

薄膜性能与其晶格常数密切相关，Maruyama 等重点研究了薄膜的电学性质与晶格常数的关系，得出：薄膜的晶格常数大于 0.3868nm 时，薄膜为导体；反之，则为绝缘体。为此，本节研究了薄膜的晶格常数与制备工艺的关系。

4.4.1 氮气分压对晶格常数的影响

图 4-12 所示为氮气分压与氮化铜晶格常数的关系曲线。由图可以看出，在开始时，随着氮气分压的增加，氮化铜的晶格常数有所增加，但随着氮气分压的继续增加到 0.8 以后，氮化铜的晶格常数随着氮气分压的增加逐渐减小。产生这种现象的原因是：氮气分压较低时，薄膜中有很多单质铜晶体，而铜的晶格常数为 0.2556nm，小于氮化铜晶格常数 0.3829nm，当薄膜的氮化铜成分增加时，薄膜的晶格常数增加；当氮气分压继续增加到 0.8 以后，薄膜中几乎没有单质铜晶体，生成的薄膜均为氮化铜，但由于刻蚀作用加强，氮气插入成键、薄膜变得紧密等原因，使得薄膜晶格常数有所减小。

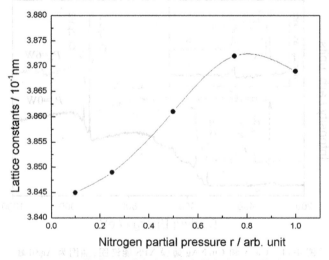

图 4-12 薄膜晶格常数与制备条件氮气分压的关系

4.4.2 掺杂对晶格常数的影响

图 4-13 所示给出了氮化铜晶格常数与薄膜内 Ti 含量的关系。由图可见,氮化铜的晶格常数随着薄膜内 Ti 含量的增加而增加,只是在中间段比较平缓,这种现象是由铜、氮化铜和氮化钛晶格常数不等所引起。

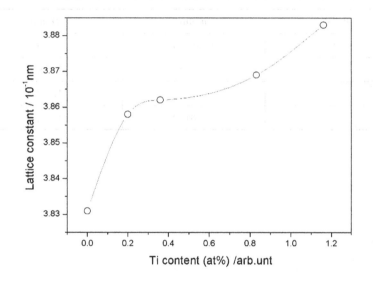

图 4-13 氮化铜晶格常数与薄膜内 Ti 含量的关系

根据薄膜的 XRD 谱,选用较强衍射峰,即小角度晶面(100)作参照,依据布拉格公式(4-2)、立方晶系面间距公式(4-3)和 Scherrer 公式(4-1)计算 Cu_3NAg 晶体的晶格常数和平均晶粒尺寸。不同 Ag 掺杂浓度 Cu_3NAg 薄膜的晶格常数 a 和晶粒尺寸 D,见表 4-4。

$$\lambda = 2 d_{hkl} \sin\theta \quad (4\text{-}2)$$

$$d_{hkl} = a/\sqrt{h^2 + k^2 + l^2} \quad (4\text{-}3)$$

式中,h、k、l 为晶面指数;a 为晶格常数;d_{hkl} 为 (hkl) 晶面族的面间距。

由表 4-4 可见,随着 Ag 的百分含量增加,Cu_3NAg 晶体的晶格常数先逐渐增大,后减小,当 $P=6W$ 时,其晶格常数最大为 0.38188nm。这与部分掺杂之后使得其晶格常数减小的报道不同[1,48],且与多数已有报道相比,在我们制备条件下制备的 Cu_3NAg,其晶格常数数值总体有些偏小[31,43,68],其原因可能是采用不同的实验制备技术所致。同时,可发现,当 $P=6W$ 时,Cu_3NAg 平均晶粒尺寸最大。

表 4-4 各掺杂功率下,Cu_3N 和 Cu_3NAg 晶体的晶格常数 a 和晶粒尺寸 D

P_w/W	a/nm	D/nm
0	0.38026	1.6369
3	0.38029	1.7824
6	0.38188	1.8485
9	0.38180	1.7788

第5章 氮化铜的性能研究

对于材料的研究,人们最终关注的还是其使用性能,如电学、光学、热学和力学性能的研究。兰州大学阎鹏勋课题组对 Cu_3N 薄膜结构开展了一系列的研究工作[23,34,37,41,65],取得了很多有价值的研究成果,探索了 Cu_3N 制备工艺(如沉积温度、氮分压、氢分压)及 Ti 掺杂等对 Cu_3N 晶体结构的影响,但其对 Cu_3N 薄膜电学性能的研究涉及不多。有研究报道:氮化铜薄膜的晶格常数大于 0.3868nm 时为导体,晶格常数小于 0.3868nm 时为绝缘体;氮化铜薄膜电阻率随晶格常数的增加有一定的增加,霍尔迁移率随晶格常数增加而增加,载流子浓度随晶格常数的增加而降低[31];Ti—Cu_3N 纳米薄膜具有明显类金属特性,且薄膜在富 N 或无游离 Ti 的情况下,带隙较大[26];Cu_3N 薄膜掺 Ti 前后,薄膜优先生长面由(111)面转向(100)面,并且薄膜的电阻率和光学带隙随着膜中 Ti 的掺杂量的增加而增大,制备过程中的氮气分压影响薄膜中的铜组分,使得薄膜具有不同的导电特性[44];而 Cu_3N 薄膜掺 La 后,薄膜的电阻率不升反降[45]。在理论计算方面,有人利用第一性原理计算发现[29],不同 Cu、N 配比及掺铅、钯体系下,Cu_3N 薄膜的电学性能可以在半导体与导体之间相互转变。并有计算发现[55],在 Cu_3N 晶格的空位中实施铜自掺杂,掺杂达 1/3 以上时,氮化铜开始由半导体转变为导体,这时氮化铜主要以 Cu_4N 和 Cu_8N 结合存在。本章详细研究了 Cu_3N 薄膜电学、光学性能及热稳定性。

5.1 薄膜的电学性能

利用四探针方法在室温下对样品的电阻率进行测量,测量结果如图 5-1 所示。可见,Cu_3N 薄膜样品的电阻率随氮气分压的增加而增大,呈指数规律变化,能够从导体的 $60\Omega \cdot m$ 变化到半导体 $400\Omega \cdot m$ 左右,然后增大至近 $5.6 \times 10^5 \Omega \cdot m$,制备的薄膜样品的电阻率仅一个勉强处于导体范围($10^{-8} \sim 10^2 \Omega \cdot m$),其余电阻率均属于半导体范围($10^3 \sim 10^7 \Omega \cdot m$)。图中显示,在较低的氮气分压下,薄膜的电阻率增长较慢,变化不大,当氮气分压较大时,电阻率则增长很快,这结果与其他研究者类似[11]。电阻率随着氮气分压按指数规律变化增加,产生这种现象的原因是:在低的氮气分压下,较多的被溅射出来单质铜原子还来不及在靶表面和基底上与氮反应,就沉积在基片上。因此,氮化铜薄膜的形成主要靠吸附氮原子插入铜原子的晶格形成 Cu—N 键生长,这时

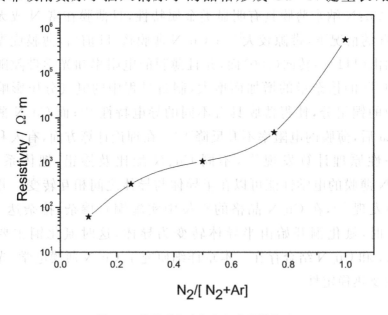

图 5-1 薄膜电阻率与氮气分压的关系

沉积在薄膜表面有相当一部分的铜原子没有与氮结合成键,所以 Cu_3N 晶格的中心空位仍含有较多的铜原子,这些填隙的 Cu 原子提供了弱局域化电子,而这些弱局域化电子相对于以共价键结合的 Cu_3N 晶格上的电子来说,具有非局域性,它的存在能改变膜内电子态密度分布,此时薄膜完全具有半导体甚至导体的特性,它的电阻率小。当氮气分压很高时,高浓度的氮氛围使得有足量的 N 原子与铜原子发生反应形成 Cu—N 键,导致薄膜中不再存在填隙的 Cu 原子,形成的电阻率相对较高,甚至接近绝缘的氮化铜薄膜。

5.2 薄膜的光学性能

5.2.1 氮分压对薄膜光学带隙的影响

图 5-2 所示是不同氮气分压 r 下沉积 Cu_3N 薄膜的 UV-VIS 透射谱(图中给出了样品相应厚度)。可见,Cu_3N 薄膜在红外区都有很好的透射率,但在不同的氮气分压 r 下有一定的变化。图谱中 a 由于沉积时氮气分压 r 小,沉积薄膜中铜相的组分较高,呈现出了与金属薄膜类似 UV-VIS 谱:在可见光区有高透过率,而在紫光和红光区附近有相对较强的吸收,形成透射峰。由于薄膜中存在 Cu_3N 相,透射峰红移,并导致在红光区的透过率比紫光区高得多。比较图谱中四条曲线发现,Cu_3N 薄膜的 UV-VIS 透射谱出现了随着氮气分压的增大,薄膜在可见光区域的透射峰逐渐减弱,而红外区的透过率逐渐加强的规律,这是由于薄膜中 Cu 相逐渐减少,Cu_3N 相逐渐增加造成的。

对于不同氮气分压 r 下沉积的薄膜,根据所得薄膜透射光谱曲线,由 Tauc 方程[68]可求得吸收系数 α 随光子能量 $h\nu$ 的变化关系,如图 5-3 所示。可见,在弱吸收区,吸收系数 α 随光子能量 $h\nu$ 基本呈指数规律上升,但在强吸收区几乎表现为线性变化,随后又有所下降。随着氮气分压 r 的增大,吸收系数 α 随光子能量 $h\upsilon$ 的变化逐渐变得平缓。

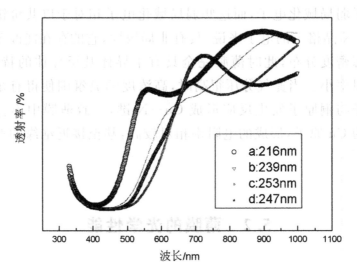

图 5-2　不同氮气分压下薄膜的 UV-VIS 谱

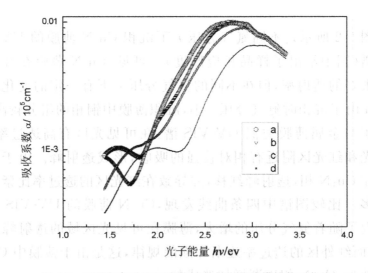

图 5-3　不同氮气分压下吸收系数与光子能量的关系曲线

根据吸收系数 α 和光子能量 $h\upsilon$ 可得到薄膜的 $(\alpha h\upsilon)^{1/2}$ 与 $h\upsilon$ 关系,将其直线部分外延到 $h\upsilon$ 轴得到了薄膜的光学带隙 E_g,如图 5-4 所示。从图中可看出,在不同的氮气分压 r 下,薄膜的光学带隙在 1.47～1.82eV 之间（与其他研究者的实验结果接近[15]）。显然,光学带隙随氮气分压 r 的增加而增大。

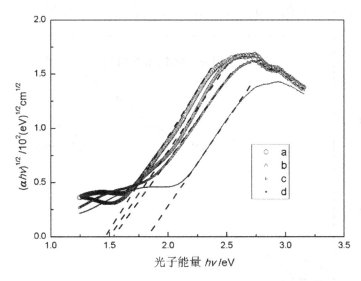

图 5-4　不同氮气分压下 $(\alpha h\nu)^{1/2}$ 与 $h\nu$ 关系曲线及光学带隙的 E_g 确定

5.2.2　掺杂对薄膜的光学带隙影响

掺 Ti 前后 Cu_3N 薄膜的 UV-VIS 谱如图 5-5 所示。从图中可以明显发现,在低波长区薄膜的透过率均比较小,在波长 600nm 以上,薄膜的透过率均比较大,但是纯 Cu_3N 薄膜在高波长区的透过率变化不大,而掺 Ti 后,薄膜的透过率随着波长的增大而增加。掺 Ti 前后 Cu_3N 薄膜的吸收系数 α 随光子能量 $h\nu$ 的变化关系,如图 5-6 所示。由图可得纯 Cu_3N 薄膜的光学带隙大约为 1.53eV,而掺 Ti 后,薄膜的光学带隙大约为 1.68 eV;图 5-7 所示为掺 Ti 的 Cu_3N 薄膜 Ti 含量与光学带隙的关系,由图可见,Cu_3N 薄膜的光学带隙随着薄膜内的 Ti 含量的增加开始时增加很快,当薄膜中 Ti 含量达到将近 20% 时,增加比较缓慢。在这里可以用 Burstein 效应来解释光学带隙增加的原因:当导带中电子浓度超过导带边缘的能态密度时,费米能级就会处在导带中,电子填充了导带底,此时由光学吸收方法所测得的带隙将会变大。另一方面,掺 Ti 后,薄膜的光学带隙变大,Ti 原子并非作为掺杂原子填入 Cu_3N 的空隙中,而是与 N 结合成 Ti—N 键。

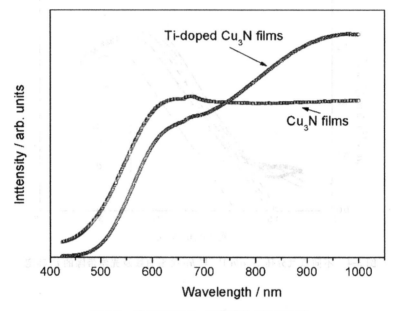

图 5-5　掺 Ti 前后 Cu_3N 薄膜的 UV-VIS 谱

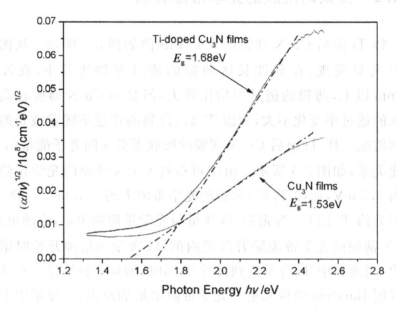

图 5-6　掺 Ti 前后薄膜的 $(\alpha h\upsilon)^{1/2}$ 与 $h\upsilon$ 关系曲线及光学带隙 E_g 的确定

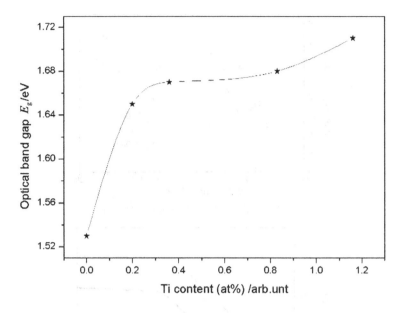

图 5-7 掺 Ti 的 Cu_3N 薄膜的 Ti 含量与光学带隙的关系

图 5-8(a)给出了 Cu_3NAg 薄膜随 P_w 变化的透射光谱,膜层相应的厚度也在图中给出。显然,在紫外至绿光波段,其透过率几乎为 0,说明绝大部分被吸收,这与我们先前的研究工作得到的结果总体上有些差异[22,44,68]。由于实验制备的 Cu_3NAg 薄膜的厚度比较大,导致其透射率相对来说比较低。Ag 掺杂后其透过率明显降低,这可能是因为 Ag 掺杂后,薄膜表面变得粗糙,对入射光产生漫反射,影响了薄膜的透过率。而当 Ag 掺杂功率为 6W 时,薄膜中的 Ag 膜对入射光的反射和吸收都增强,加上更加粗糙的薄膜表面粗糙度,使得 Cu_3NAg 薄膜的透过率降低得更快。

由透射光谱结合 Urbach 模型[115]以及 Tauc 模型[57,116],通过计算可获得薄膜的 Urbach 能量 E_u 以及光学带隙 E_g。E_u 反映带尾态密度随能量指数分布的斜率,即带尾态宽度,E_u 的变化与薄膜内缺陷密度变化一致,E_u 大薄膜内缺陷密度大,E_u 小薄膜内缺陷密度小。计算 E_u 与 E_g 可由公式(5-1)和(5-2)确定:

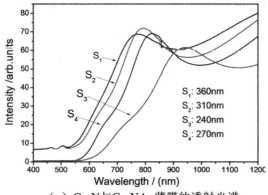

（a）Cu_3N 与 Cu_3NAg 薄膜的透射光谱

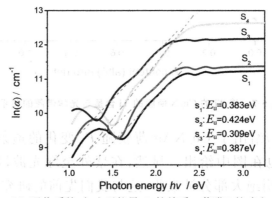

（b）吸收系数α与光子能量 $h\nu$ 的关系，薄膜 E_u 的确定

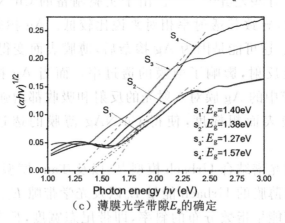

（c）薄膜光学带隙 E_g 的确定

图 5-8　Cu_3NAg 薄膜随 P_w 变化的透射光谱

$$\alpha h\nu = A\exp(-h\nu / E_u) \quad (5\text{-}1)$$

$$(\alpha h\nu)^{1/2} = B(h\nu - E_g) \quad (5\text{-}2)$$

式中，A_0、B 均为常数，$h\upsilon$ 为光子能量。吸收系数 α 由公式(5-3)给出：

$$\alpha = \ln[100/T]/d \tag{5-3}$$

式中，T 为透过率，d 为薄膜的厚度。

图 5-8(b) 为 Cu_3NAg 薄膜吸收系数 α 与光子能量 $h\upsilon$ 的关系，E_u 为其线性部分斜率的倒数。图 5-8(c) 为计算的 $(\alpha h\upsilon)^{1/2}$ 与光子能量 $h\upsilon$ 的关系图，线性外推可得 E_g。显然，在 P_w 为 6W 时，Cu_3NAg 薄膜的 E_u 最小，说明此条件下沉积的薄膜结晶好，内部的缺陷密度最少，这正好和前面 SEM 结果相符合。Cu_3NAg 薄膜的 E_g 随 P_w 的增大先缓慢减小，当 P_w 为 6W 时，其 E_g 为 1.27eV。因此，可以通过改变 Ag 掺杂浓度，实现对 Cu_3NAg 薄膜 E_g 的合理控制。

对 Cu_3N 进行掺杂能够有效地控制各种缺陷能级的浓度，进而调控薄膜的可见光发射。Cu_3NAg 薄膜可见光区的发光主要是由薄膜中的缺陷、杂质浓度决定的，Cu_3NAg 薄膜内缺陷主要有晶格空位、银填隙和银取代铜等。我们对 Cu_3NAg 薄膜进行了室温光致发光测试，结果如图 5-9 所示。当激发波长为 310nm 时[图 5-9(a)]，Cu_3N 和 Cu_3NAg 薄膜在蓝光区域都出现了 490nm 和 522nm 两个峰，随着膜中银浓度的增加，这两个峰衰减较快。当激发波长为 370nm 时[图 5-9(b)]，薄膜在蓝光区域都出现了 486nm 和 506nm 两个峰，同时，在紫光区域 418nm 处出现了一个很强的峰，随着膜中银浓度的增加，这些峰位明显加强。通过光致发光的发光峰位，容易得到 Cu_3N 和 Cu_3NAg 薄膜的本征发光带隙大致在 2.38~2.97eV。这一结果比前面计算所得的 E_g 大 0.81eV 以上，说明这些光致发光是薄膜内的深能级缺陷产生的。由发光光谱结合 EDS 结果，我们认为，随着 P_w 的增大，薄膜内银的浓度随之增加，更多的银原子掺入薄膜内部，并占住 Cu_3N 晶格内空隙位，薄膜的浅能级缺陷逐渐向深能级转变，当薄膜内银浓度为 0.15at% 时达到最高。空隙占位的银原子产生了更多深能级杂质复合中心，减短薄膜内非平衡载流子寿命，使得薄膜的电子输运性减弱。光致发光结果证实 Ag 元素的掺杂能够有效改变 Cu_3N 薄膜内深能级缺陷。

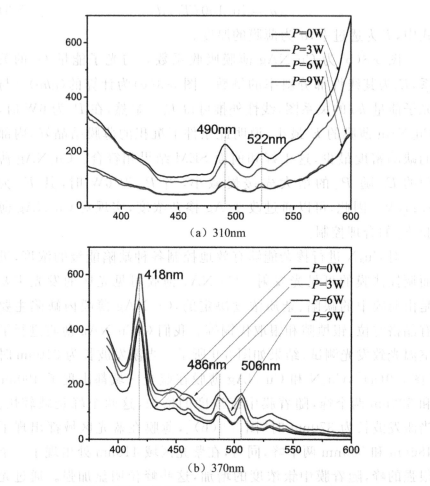

图 5-9 不同激发光波长 Cu_3N 和 Cu_3NAg 薄膜的 PL 谱

5.3 薄膜的热稳定性研究

热重分析(TG)是在一定的气氛中,测量样品的质量随温度或时间变化而变化的技术,利用此技术可以研究诸如挥发或降解等伴随有质量变化的过程。图 5-10 所示是 N 流量为 0.1 和 0.75、射频功率为 150W、沉积温度为 100℃时,沉积薄膜的热失重(TG)曲线(实验样品的总质量分别为 1.413mg 和 1.602mg)。由图可见,在 N 流量为 0.1 和 0.75 条件下制备的样品,其分解温度分别约在 547K 和 516K,这些分解比已有的报道偏低。在 TG 曲线中,薄膜在热解过程中出现了 3 个明显的阶段:第一阶段从 300K 到 530K,此阶段温度范围变化较大,但是薄膜的失重却很小,这个失重主要是膜在空气中吸附的水分和一些不稳定的悬挂键从膜中解析出来所致;第二阶段从 530K 至 650K,这个阶段热失重曲线变化比较大,薄膜的失重量也最大,此时膜内的 Gu—N 键已经发生断裂,第三阶段在 650K 以上,在这个温度以后薄膜大部分已经分解,薄膜中大部分 N 已经脱离,因此失重逐渐减小,直到全部剩下铜元素为止。

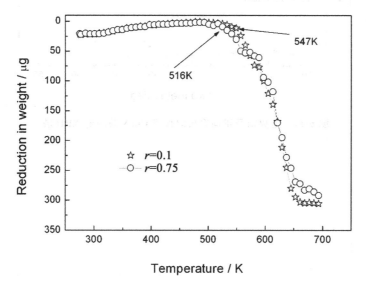

图 5-10　不同氮流量下制备的 Cu_3N 薄膜的 TG 曲线

图 5-11 所示是 Cu_3N 在 473K 和 573K 温度下退火前后的 XRD 谱。由图可见,在退火前在 XRD 谱中主要有对应于 Cu_3N 四个峰(100)、(110)、(111)和(200)。在 473K 退火温度下薄膜的 XRD 谱中出现了铜(111)峰,与此同时,Cu_3N (111)和(200)峰略微减小。当退火温度达到 573K 后,XRD 谱中的铜(111)峰变得很强,同时也出现了铜(200)峰。而 Cu_3N (111)和(200)峰几乎消

失,说明薄膜中 Cu_3N 相几乎全部转化为 Cu 相,也就是说,Cu_3N 薄膜在 573K 温度下基本全部分解为铜和氮气。这一结果与前面 TG 得到的结果基本相符。

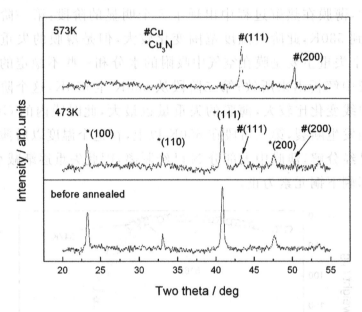

图 5-11 退火前及不同退火温度下 Cu_3N 薄膜的 XRD 谱

第6章 氮化铜的第一性原理研究

6.1 概 述

21世纪是尖端材料科技与生物科技的时代,而尖端材料与有机分子的微观结构对了解其电子结构、光学性质、温度的影响、磁的特性、机械的特性等等是不可或缺的。20世纪初期由于量子力学的发现,使我们有机会从微观的角度去探讨以上的问题,藉由第一原理的材料计算方法求出其所有物性。对于实际的物理问题,往往很难得到解析解答。在计算机发明之前和计算机功能还不是很强大时,理论物理学家必须将实际的物理问题简化成可以用数学解析的简单模型。这种方式可以获得定性上的结果。如果模型选得对,也可以得到重要的物理知识。但是如果模型选择不正确,可能误导以后的物理研究以及产生不正确的物理观念。随着计算机的发明及计算机功能的不断强大,很多不能单靠数学得到解析的物理问题,可以利用计算机数值的计算得到答案,理论计算物理因此开始发展。

由于实验研究往往需要用到昂贵的实验仪器以及材料,有时候实验的样品也很不容易获得,更有许多实验需要用到高度污染或有剧毒的药品或材料。这些实验如用理论计算模拟来取代,将可以节省很多研究经费,减少许多环境的污染,以及缩短得到答案及结果的时间,因为实验研究没有理论研究的配合,往往采用"试错法(Try and error)",即猜—试—错了—再猜试的步骤。但理论计算的模拟可以作为前期评估并将实际研究导入正确的方向。所以,随着计算机功能的急速增加,理论计算研究将越来越重要。但目前理论计算物理、计算材料科学的应用仍受到计算机功能的限制,尤其是使用物理基本原理的计算方法。能够使用的物理体系模型仍与实际的物理体系有相当大的差别,例如半导体数百至数千原子厚的磊晶,触媒里半径约为 $10^{-2}\mu m$ 的白金或铑的金属稞粒等。但现在能够研究的问题比起二十年前的已经增加了很多。例如使用高度平行处理的多处理机计算机已经可以计算硅(111)7×7 这么复杂的表面。当然,实验研究还是必需的,尤其是在最后一步要进入实际应用的阶段。

计算物理学是一门交叉学科,是物理学、数学、材料科学、计算机科学等学科相结合的产物,它是这几门学科高度发展的结晶,它是以计算机及其技术为工具和手段,解决复杂物理问题的一门应用性学科。计算材料学是以计算机为手段,通过理论与计算对材料的固有性质、结构与组分、使用性能以及合成与加工等综合研究的一门新兴学科,材料计算已经成为现代科学研究中一个极具前景且最为活跃的领域之一,对材料科学的发展起着极大地推动作用。这主要是因为:其一,材料计算可以对材料进行设计,具有前瞻性;其二,它能在计算机计算能力许可的范围内进行创新探索,具有创新性;其三,它可进行模拟实验,减少或代替部分实验工作。近年来,采用"第一性原理(First-principles)"计算研究材料的结构和性能,已经成为材料学科的研究工作中一个不可或缺的环节。随着计算机功能的强大以及计算方法的不断改进,可以预期在不久的将来,可靠性高的物理基本原理计算方法将能够使用很接近实际的物理系统的模型,因此能更真确地模拟实际的物理现象。

所谓"第一性原理"是指在计算过程中不需要由实验提供参数,只要知道材料组成的元素便可直接求解其对应的薛定鄂方程,求出其所有的物性。但由于这是一个多电子的问题,处理起来非常困难,直到 20 世纪 60 年代 Walter Kohn 教授及沈吕九教授提出局部密度泛函近似理论(LDA)才使这个沉潜多年的问题重露曙光,经过多年计算机仿真计算的验证,LDA能对非强关性系统提供一个非常好的基态描述,而随着高速计算机效能的日新月异,更使第一性原理材料计算方法可普遍应用于绝大部分的实验研究系统。

6.2 计算方法及过程

半导体、非线性光学材料、金属氧化物、玻璃、陶瓷等固体材料,对电子工业、航空航天以及石化、化工等工业领域有着非常重要的战略意义。对这些材料而言,其电子的结构与性质,以及表面和界面的性质与行为都非常重要。半导体和其他固体材料的许多性能由电子性质决定,而电子性质又由原子结构决定,特别是缺陷在改变电子结构上的作用对半导体性质尤为重要。分子模拟,特别是量子物理技术,可用来预测原子和电子结构及分析缺陷对材料性能的影响。CASTEP能有效地研究存在点缺陷、空位、替代杂质、位错等的半导体和其他材料中的性能。CASTEP的量子力学方法,为深入了解固体材料的这些性质以及设计新的材料,提供了强有力的工具。

材料计算的过程是根据材料的本身构成(结构和组分)及基本物理理论,建立物理模型,继而根据具体的模型,建立势模型并构造势函数。

6.2.1 CASTEP软件介绍

1. CASTEP软件的主要功能

基于密度泛函平面波赝势方法的CASTEP软件可以对许多体系包括半导体、陶瓷、金属、矿石、沸石等进行第一原理量子力学计算。典型的功能包括研究表面化学、带结构、态密度、和光学性质。它也能够研究体系电荷密度的空间分布和体系波函数。

CASTEP 还可以用来计算晶体的弹性模量和相关的机械性能,如泊松系数等。CASTEP 中的过度态搜索工具提供了研究气相或者材料表面化学反应的技术。

CASTEP 可以实现:计算体系的总能;进行结构优化;执行动力学任务;在设置的温度和关联参数下,研究体系中原子的运动行为;计算周期体系的弹性常数;化学反应的过度态搜索等。除此之外,计算一些晶体的性质,如能带结构、态密度、聚居数分析、声子色散关系、声子太密度、光学性质、应力等。量子力学计算精确度高但计算密集。直到最近,表征固体和表面所需的扩展体系的量子力学模拟对大多数研究者来说才切实可行。然而,不断发展的计算机功能和算法的进步使这种计算越来越容易实现。

2. CASTEP 软件的主要理论

(1)密度泛函理论(DFT)。CASTEP 的理论基础是电荷密度泛函理论在局域电荷密度近似(LDA)或是广义梯度近似(GGA)的版本。DFT 所描述的电子气体交互作用被认为对大部分的状况都足够精确,并且它是唯一能实际有效分析周期性系统的理论方法。

Hohenberg-Kohn 理论:体系的电子行为由 Schrodinger 方程描述。如果只考虑系统的平衡态,则电子结构与时间无关,由定态 Schrodinger 方程描述:

$$H\psi = E\psi \qquad (6-1)$$

式中,E 为电子的能量,$\psi = \psi(X_1, X_2, \cdots\cdots, X_N)$ 是多电子波函数(X_i 为电子 i 的空间坐标和自旋坐标),H 为哈密度算符。在由原子组成的体系中,由于原子核比电子的质量大得多($10^3 \sim 10^5$ 倍),因此在研究电子结构时,可以认为原子核固定不动,这就是所谓的 Born-Oppenhermer 近似(或称绝热近似)。对于超过两个电子以上的体系,Schrodinger 方程(6-1)是很难于严格求解的,因此从 Schrodinger 方程更不能严格求解多电子体系的电子结构。而密度泛函理论将多电子波函数 $\psi(X_1, \cdots\cdots, X_N)$ 和 Schrodinger 方程用非常简单的电荷密度 ρ 和对应的计算方案来代替,提供了一条研究多电子系统的电子结构的有效途径。

1964 年,Hohenberg 和 Kohn 建立起密度泛函理论的基本框架。首先采用电荷密度 ρ 作为描述体系性质的基本变量并提出了两个定理。第一定理表述为:外场势是电荷密度的单值函数(可相差一常数),它的推论是,任何一个多电子体系的基态总能量都是电荷密度 ρ 的唯一泛函,ρ 唯一确定了体系的(非简并)基态性质。第二定理表述为:对任何一个多电子体系,总能的电荷密度泛函的最小值为基态能量,对应的电荷密度为该体系的基态电荷密度。Hohenberg-Kohn 的密度泛函理论只有对基态才是严格成立的,这使得将 DFT 应用在考虑电子作用的核动力学的计算中受到一定的限制。

局域(自旋)密度近似:它在第一性原理计算中得到了广泛的应用,并且在大多数情况下给出了较好的结果,与实验结果符合的很好,然而在某些方面还存在不足。严格地说局域密度近似只适用于密度变化足够缓慢或者高密度情况,对于一般的密度变化并不缓慢的体系的描述,理论上并不清楚。在计算上,人们对局域密度近似进行了改进和修正,比如自相互作用修正(SIC)、自能修正(SEC 或 GWA)、在为库仑修正[L(S)DA+U]以及广义的梯度近似(GGA)。

在 CASTEP 里预设的是 GGA,在很多状况下它被认为是比较好的方法。LDA 会低估分子的键长(或键能)以及晶体的晶格参数,而 GGA 通常会补救这缺点。梯度修正的方法在研究表面的过程、小分子的性质、氢键晶体以及有内部空间的晶体(费时)是比较精确的。有许多证据显示 GGA 会在离子晶体过度修正LDA 结果;当 LDA 与实验符合得非常好的时候,GGA 会高估晶格长度。

(2)赝势。电子-离子间的交互作用可以用赝势的观念来描述。CASTEP 中有两种赝势,一种是规范-守恒赝势(Norm-conserving pseudopetential),另一种是超软赝势(ultrasoft pseudopotential)。

　　Norm-conserving 赝势是相当有名的而且是经彻底验证的。在这种方法中，赝波函数在定义的核心区域的截止半径以上是符合全电子波函数的。它要求改造后的波函数其在截止半径 R_c 之内的总电荷量仍要等于未改造前 R_c 之内总量的大小，这样赝势的精确度能够大幅度地提升。因此，我们取距原子中心 R_c 处为划分点，赝势产生示意图 R_c 以上波函数完全一样保留，而 R_c 以内则对波函数加以改造。主要是要把振荡剧烈的波函数改造成一个变化缓慢的波函数，而它需要是没有节点的。少了剧烈振荡不但允许只以相对很少的平面波来展开波函数，没有节点的（径向）波函数也意味着没有比它本征值更低的量子态来与它正交。求解内层电子的需要就自动消失了。以这样一个假的赝势能够在同样的本征值的情况下给出一价电子近似解，所以把它叫做是赝势 Vpseudo(Vp)。在 CASTEP 中引用的是最佳化的方法，然而描述第一列（碳、氮、氧）或过渡金属（镍、铜、钯）等局域化价电子轨域的所需截止能量仍然太高。norm-conserving 赝势能够在实空间或是倒空间的波函数来使用，实空间的方法提供了对于系统而言比较好的可测量性。

　　超软赝势（ultrasoft pseudopotential）的特色是让波函数变得更平滑，也就是所需的平面波基底函数更少。Vanderbilt 所提出来的超软赝势的想法是不用释放非收敛性条件，用这样的方法来产生更软的赝势。在这个方法里，虚波函数在核心范围是被允许作成尽可能平滑的，以致于截止能量可以被大大地减小。就技术上而言，这是靠着广义的正交条件来达成的。为了要重建整个总的电子密度，波函数平方所得到电荷密度必须是核心范围再加以附加额外的密度进去。这个电子云密度因此就被分成两部分，第一部分是延伸至整个单位晶胞的平滑部分，第二部分是在核心区域局域化的自旋部分。前面所提的附加部分只出现在密度，并不在波函数。这和像 LAPW 那样的方法不同，在那些方法中类似的方式是运用到波函数。

超软赝势产生算法保证了在预先选择的能量范围内会有良好的散射性质,这导致了赝势更好的转换性与精确性。超软赝势通常把每个角动量通道当作价电子来处理浅的内层电子态,这也会使精确度和转换性进一步提升,虽然计算代价会比较高。目前,超软赝势只可以在倒置空间中使用。

(3)分子轨道的自洽求解。密度泛函理论是基于 Hohenberg-Kohn 定理,该定理表明体系基态的性质由电荷密度决定,体系的总能量是电荷密度 ρ 的函数。总能 E_t 可以表达为:

$$E_t[\rho] = T[\rho] + U[\rho] + E_{xc}[\rho]$$

$T[\rho]$ 是密度为 ρ 的电子的动能,$U[\rho]$ 是经典的库仑相互作用静电能,$E_{xc}[\rho]$ 包括了多体相互作用对总能的贡献,其中交换-关联能是主要的部分。从波函数构造电子密度,最终得到分子轨道的自洽场方程:

$$HC = \varepsilon SC$$

$$H_{\mu\nu} = \langle \chi_\mu(r_1) | \frac{-\nabla^2}{2} - V_N + V_e + \mu_{xc}\rho \, r_1 | \chi_V(r_1) \rangle$$

$$S_{\mu\nu} = \langle \chi_\mu(r_1) | \chi_\nu(r_1) \rangle$$

它是非线性方程,只能用迭代方法求解。由给定初始的 C_{iu},构造初始的分子轨道 φ,再构造电荷密度,然后计算出 H,代入 $HC=\varepsilon SC$ 求出新的 C_{iu},计算新的 φ 和新的 ρ_{in},若 $\rho_{in} = \rho_{out}$,则可计算出总能 E_t,进一步得到其他性质。

3. 主要设置

(1)计算任务的设置。在 CASTEP 软件中进行任务设置,主要是通过 Visualizer 应用窗口中的工具条之一"Calculation"来进行。我们可以更改工具框中的相应选项,来配置诸如"电子选项""结构优化选项"和"电子和结构性质选项"等。这几个选项是我们在运用 CASTEP 计算研究中非常重要的几个技术参数。其中,"电子选项"是很多其他计算任务也要涉及的。在 CASTEP 中还有如动力学、结构优化、弹性常数、过渡态等计算的设置。在程序运行之前,从研究的问题出发,要将软件中一些关键的任务参数设置成符合计算需要的值,我们才能得到所期望的运算结果。在利用 CASTEP 做有关能量、动力学、结构优化、弹性常数、过渡态等计算时,必须对电子选项进行设置。在电子选项中主要有精度设置、交换-关联函数的设置、赝势的设置、截断能的设置、K 点的设置。

(2)结构优化任务的设置。结构优化是 CASTEP 计算中经常要进行的计算任务,特别是想要计算所关注体系的各种性质的时候,必须首先进行结构优化的计算,在得到结构优化结果文件以后,才能进行性质的计算。所以,正确地设置结构优化的参数是非常重要的。在 CASTEP 软件中,有四个来控制结构优化的收敛参数:第一个是能量的收敛精度,单位为 eV/atom,是体系中每个原子的能量值;第二个是作用在每个原子上的最大力收敛精度;第三个是最大应变收敛精度,单位为 GPa;第四个是最大位移收敛精度。这些收敛精度指的是两次迭代求解之间的差,只有当某次计算的值与上一次计算的值相比小于设置的值时,计算才停止。

(3)计算体系性质的设置。在 CASTEP 中可以计算体系的性质,如能带结构、态密度、聚居数分析、声子色散关系、声子态密度、光学性质、应力等。在进行能带和态密度这两项的计算设置之前,必需先进行自洽计算得到基态能量,而结构优化能够做到这一点,所以要在计算能带和态密度之前对体系进行结构优化。

(4)计算结果的分析。如果计算时把计算模型取名为 66,能带计算完成后,会有名为 *.castep 文件生成。首先在 Visualizer 界面中把该文件打开,接着点击 Visualizer 应用窗口中的工具条"Analysis"就会有对话框出现。该对话框中的"Scissors"选项是剪刀工具,可以把能带作一个微调。选择图形显示的方式,分为点、线、点线结合三种。若选择线"Line",在计算能带以后,可以同时把总的态密度显示出来。然后选中"view"按钮,则在 Visualizer 界面中会显示能带和对应的总的态密度图,得到的能带和总的态密度图还可以导出到如 origin 软件中进行处理,以利于更直观的分析。

6.2.2 模型的构建及计算

氮化铜属于反三氧化铼结构,空间结构群属于:Pm3m,晶格常数约为 3.85nm,其中氮原子占位(0,0,0),铜原子占位为(1/2,1/2,1/2)。利用 CASTEP 软件构建氮化铜空间结构图,如图 6-1 所示。

计算采用基于第一性原理密度泛函理论结合平面波赝势方法的 CASTEP 软件包。计算过程中,采用周期性边界条件,电子间的交换关联采用广义梯度近似(GGA-PBE)方法,由此,电子波函数则利用平面波基矢扩展,而且采用模守恒赝势描述电子-离子相互作用。截止能取 300eV,能量计算都在倒易空间中进行,获得优化晶体结构,在此基础上进一步进行电子结构和态密度的计算。

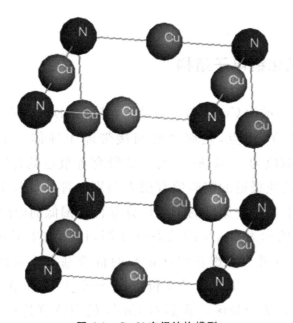

图 6-1　Cu_3N 空间结构模型

6.3 氮化铜的电子结构计算结果

固体的导电性能是由其能带结构决定的。对一价金属来说，其价带是未满带，故能导电。对二价金属来说，价带是满带，但其禁带宽度为零，价带与其较高的空带相交叠，满带中的电子能占据空带，因而二价金属也能导电，半导体和绝缘体的能带结构相似，价带为满带，价带与空带间存在有禁带。半导体的禁带宽度在 $0.1 \sim 4eV$ 之间，绝缘体的禁带宽度在 $4 \sim 7eV$ 之间。在任何温度下，由于热运动，满带中总会有部分电子具有足够的能量激发到空带中，使之成为导带。由于绝缘体的禁带宽度较大，通常情况下，从满带激发到空带的电子数微不足道，一般表现为导电性能差。半导体由于其禁带宽度较小，满带中的电子只需要较小能量就能激发到空带中，半导体在宏观上表现为有较高的电导率。

6.3.1 氮化铜电子结构

图 6-2 所示为 Cu_3N 的能带结构，其价带顶位置在 R 点，而其导带底位置在 M 点，是典型的间接带隙半导体。本文计算的 Cu_3N 的带隙值为 $0.3218eV$，与其他研究者报道的计算值相差不多，但是与其他报道的实验值有很大差别。理论计算值小于实验值，这是由于第一性原理的带隙计算都普遍偏低的原因所致。

图 6-3 所示为 Cu_3N 的总态密度图，计算结果显示 Cu_3N 是一种半导体材料。不难发现，Cu_3N 的价带基本上可以分为三个区域，即 $-8.0 \sim -5.0eV$ 的下价带区、$-5.0 \sim 0eV$ 的上价带区，以及 $-16eV$ 的价带区。我们对 Cu_3N 的态密度进行分析，做出原子外层电子分轨道态密度，如图 6-4 所示。铜是第 29 号元素，铜的电子排布式是：$1s^2\backslash\backslash 2s^2\backslash\backslash sp6\backslash\backslash 3s^2\backslash\backslash 3p^6\backslash\backslash 4s^1\backslash\backslash 3d^{10}$，这里

由于洪特规则特例,要符合能量最低原理(当轨道处于全空、半空、全充满的时候能量较低,原子较稳定),所以 4s 上的一个电子被激发到 3d 轨道上去,而激发的能量可以通过满足 $4s^1$ 的半充满和 $3d^{10}$ 的全充满来弥补,这样使得铜原子的能量最低。

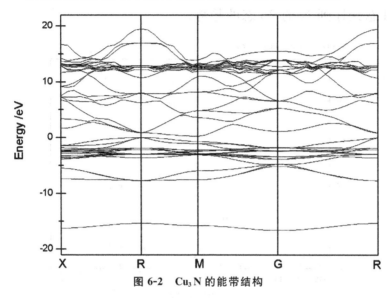

图 6-2　Cu_3N 的能带结构

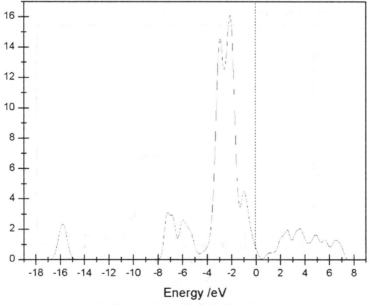

图 6-3　Cu_3N 的总态密度图

从图 6-4(a),(b)中可以看出,对上价带区贡献最大的是铜的 3d 态电子,N 的外层 2p 态电子对下价带区也有点贡献;而 −16eV 的价带区部分,是由 N 的 2s 态及铜的 s 和 p 态形成的。对于导带部分,其主要来源于铜的 3d 态的贡献,且电子具有从铜的 3d 态跃迁到 N 的 2p 态的趋势。

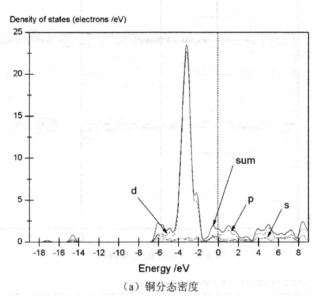

(a) 铜分态密度

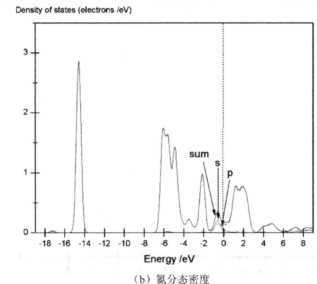

(b) 氮分态密度

图 6-4　Cu_3N 原子外层电子分轨道态密度

6.3.2 掺杂氮化铜电子结构

掺杂 Cu_3N 晶体结构模型,杂质位于空位中心,如图 6-5 所示(以铅为例),优化计算后,得到的晶胞参数有所增大。

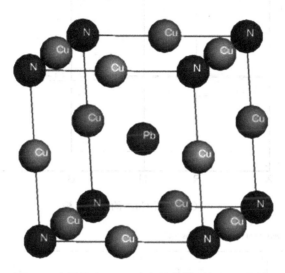

图 6-5 金属掺杂 Cu_3N 空间结构模型(以 Pb 为例)

图 6-6(a)、(b)分别为掺铅 Cu_3N 的能带结构和原子外层电子分轨道态密度。从两个图中不难看出,掺铅 Cu_3N 明显具有导电性能,而其导电性能主要来自 p 和 d 轨道的电子,因为铅的外层电子组态为 $4f^{14}\backslash\backslash5d^{10}\backslash\backslash6s^2\backslash\backslash6p^2$,铜的外层为 $4s^1\backslash\backslash3d^{10}$。比较图 6-6(a)与图 6-2 可以看出,掺铅之后,Cu_3N 的价带向下移,Cu_3N 的带隙在掺杂前处于费米面的上方,掺铅后带隙消失,Cu_3N 由半导体转变为导体。

图 6-7(a)、(b)分别为掺镧 Cu_3N 的能带结构和原子外层电子分态密度。由图可以看出,掺镧 Cu_3N 也明显具有导电性能,而其对电导率贡献的电子与掺铅有点不同,其主要来自 d 轨道的电子,其次是 p 轨道的电子。因为镧的外层电子组态为 $5d^1\backslash\backslash6s^2$,又由于铜的 3d 电子,因此,其电导率主要来自 d 轨道电子的贡献。

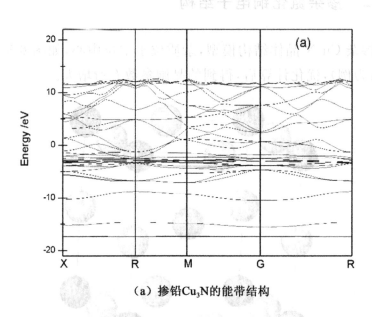

（a）掺铅Cu_3N的能带结构

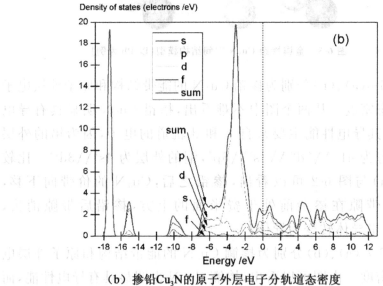

（b）掺铅Cu_3N的原子外层电子分轨道态密度

图 6-6　掺铅 Cu_3N 的能带结构和原子外层电子分轨道态密度

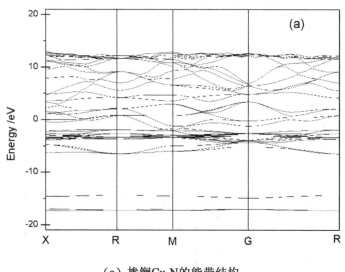

(a) 掺镧Cu_3N的能带结构

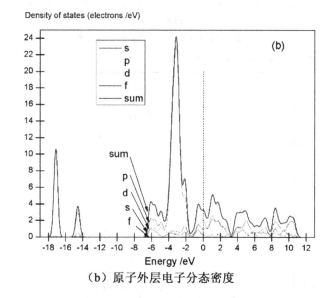

(b) 原子外层电子分态密度

图 6-7 掺镧 Cu_3N 的能带结构和原子外层电子分态密度

图 6-8 (a)、(b)分别为掺钛 Cu_3N 的原子外层电子分态密度和钛原子分态密度图。由图可以看出,掺钛 Cu_3N 导电性比掺铅和镧更加明显,比较图 6-8 (a)、(b)不难得出,其导电性主要来自 Ti 原子的贡献。由图 6-8(b)可见,掺钛 Cu_3N 的导电性能其主要

来自钛 d 轨道的电子,而 s 和 p 轨道的电子对导电性的贡献则相对较小。

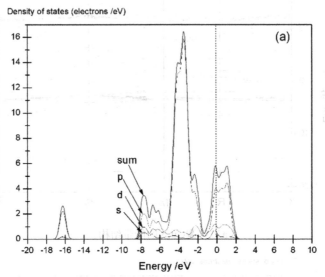

（a）掺钛Cu_3N的原子外层电子分态密度

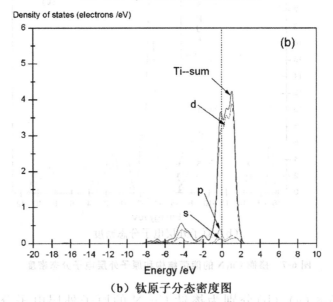

（b）钛原子分态密度图

图 6-8　掺钛 Cu_3N 的原子外层电子分态密度和钛原子分态密度图

6.3.3 态密度分析小结

氮化铜掺杂后,其晶格常数有所增大,使晶胞扩大。掺杂氮化铜的总体态密度和未掺杂氮化铜的总体态密度变化较大。掺铅之后,价带态密度变窄,导带态密度展宽,并且在-10.0eV附近和-17.0eV附近分别出现了一个一弱一强的附加态,这些态均由于掺杂铅元素的p态形成;掺镧后,态密度总体变化与掺铅类似,在-10.0eV附近没有附加态出现,而在-17.0eV附近出现稍弱一个附加态,这些态均由于掺杂铅元素的d态形成;氮化铜掺钛后,导电性的变化比掺铅和镧更大,这一变化主要来自钛原子的掺入。氮化铜掺杂前后相比,掺杂后,氮化铜的态密度均向低能方向移动,费米能级进入导带,说明导带底有大量的电子存在,氮化铜的导电性能明显增强。

结论与展望

本书首先简略介绍了薄膜材料的应用及分类,介绍了目前制备薄膜的常用技术,然后比较详细地介绍磁控溅射制备薄膜技术及与氮化铜薄膜有关的相应测试分析技术。在详细综述了氮化铜研究现状的基础上,采用射频磁控溅射方法在不同工艺条件下制备了一系列氮化铜薄膜(掺杂)样品,采用场发射扫描电子显微镜(SEM)、原子力显微镜(AFM)、X射线光衍射仪(XRD)、X射线光电子能谱仪(XPS)、紫外-可见光光谱仪(UV-VIS)、荧光磷光光度计、热重分析仪(TG)、霍尔效应测试仪和四探针测试仪等对样品进行了表征分析,研究了样品结构、性能及制备工艺条件之间的关系,并结合第一性原理深入探索了氮化铜的电子结构及电学性能,研究发现:

(1)Cu_3N薄膜结构受性能受制备工艺条件的影响,沉积功率、温度及氮气分压对其影响较大,重点研究了氮气分压与薄膜生长及电学性能的关系。研究发现,薄膜呈现由低氮气分压的Cu_3N(111)晶面转向高分压的(100)面择优生长规律,而沉积温度对薄膜的择优生长有一定的影响,但沉积功率只能改变薄膜的生长速度。在性质方面,制备的薄膜由工艺条件的不同分别表现为半导体和绝缘体的电学性质。

(2)在Cu_3N薄膜的结构、组分及掺杂对薄膜性能影响的实验研究中,主要研究了掺入不同浓度的Ti原子对薄膜性能的影响。掺Ti后,薄膜的晶格常数增大,表面变得光滑均匀,且外形由金字塔形变为半球形。通过XRD谱发现,Ti原子与N形成Ti—N键,随着Ti含量增加,Ti—N增强,并且薄膜的定向生长随之变化;四探针测试结构发现,掺 Ti 后,氮化铜的电阻率在 Ti 含量浓度较低时(小于 0.2at%)增加较快,但薄膜仍是半导体材料特性;通过 UV-VIS 结果结合 Tauc 方程,得到了薄膜的光学带隙随着薄膜内 Ti 含量的增加而增大。Ag 的变化对 Cu_3N 薄膜表面生长方式影响很大。在银掺杂实验研究中发现,当 Ag 溅射功率较小时,Cu_3N 薄膜表面光滑,晶粒尺寸较小;随着 Ag 溅射功率的增大,薄膜内 Ag 含量增加,薄膜的表面变得粗糙,晶粒尺寸变大,然后又有所减小。随着 Ag 浓度的增加,Ag 原子选择填入 Cu_3N 晶格中空位,薄膜的光学带隙先降后升,薄膜内能级缺陷向深能级方向,薄膜的电子输运性变弱。表明掺 Ag 可以有效调控 Cu_3N 膜的晶体结构、能级结构和光学性能,使其更满足光电材料及器件的需要。

(3)利用基于密度泛函理论的第一性原理建立 Cu_3N 结构模型并对模型进行优化,进而研究 Cu_3N 能带结构、态密度等电子学性能。在电子结构图中发现,纯氮化铜上价带区贡献最大的是铜的 3d 态电子,其带隙大约为 0.3218eV;掺入杂质 Ti、Pb、La 后,氮化铜的电子结构和态密度与纯氮化铜相比均发生了较大的变化,但其变化趋势基本雷同:费米能级进入导带,带隙消失,且氮化铜的态密度均向低能方向移动,氮化铜薄膜具有导体特性。

总之,氮化铜薄膜的制备方法较多,不同的制备工艺对 Cu_3N 薄膜的结构和性能影响很大,不同研究者报道的研究成果有的大相径庭。相对而言,磁控溅射法制备 Cu_3N 薄膜技术比较稳定,基本得到研究者的认可,但是由于利用磁控溅射技术制备 Cu_3N 薄膜其可控参数较多,制备参数改变对薄膜的结构和性能产生的影响还不是十分明了。

迄今为止,Cu_3N 薄膜制备工艺探索、薄膜的结构和电学、光学性能方面开展的研究工作较多,取得了一定的成果,但在 Cu_3N 薄膜的应用方面开展的研究工作还不是很多。今后可以在以下应用方面开展研究工作:

(1)Cu_3N 薄膜制备简单、无毒,可取代现有一次性光存储碟基无机相变材料。

(2)Cu_3N 薄膜在较低温下热分解得到铜膜,有望作为集成电路中金属 Cu 线的缓冲层、低磁阻隧道结的障碍层、自组装材料的模板等。

(3)Cu_3N 薄膜具有良好的电子发射特性,是场发射材料的有力竞争者。

(4)Cu_3N 颗粒分解物可提供良好传导电子的性能,它有望应用于新能源储能材料(锂电池负极材料)。

Cu_3N 薄膜结构独特,性能优异,有望成为新一代的电学、光学和热学等领域功能材料。现阶段,使用磁控溅射制备 Cu_3N 的方法基本得到一致认可,薄膜的生长速率、薄膜晶体结构随制备工艺的变化而改变的研究已取得了不错的成效,对未掺杂 Cu_3N 薄膜的电学、光学性质的研究也取得了一定的成果,但薄膜掺杂后所引起其电、光学性能的改变的研究在国内外才刚刚开始,需进一步的深入。Cu_3N 薄膜掺杂后,其性能有很大的变化,但人们对其作用机理、变化规律、掺杂原子的占位情况,以及最佳掺杂元素及浓度等问题都还不十分明了,特别是 Cu_3N 晶体结构中,掺杂原子壳层结构及内外层电子数的关系几乎没有研究,而这又是影响 Cu_3N 性能的关键因素。另外,如何通过改善制备工艺条件,调制 Cu_3N 薄膜的组分与结构,从而使其分解温度满足特定的工作条件,增强 Cu_3N 薄膜材料工作稳定性,并简化其与制备工艺之间的矛盾,这些工作还有待进一步的深入开展。如果上述问题可以得到圆满解决,Cu_3N 薄膜应用于先进工业技术材料、促进经济的快速发展、造福人类社会将指日可待。

参考文献

[1] A. A. Yu, Y. H. Ma, A. S. Chen, et al. Thermal stability and optical properties of Sc-doped copper nitride films[J]. Vacuum, 2017, 141: 243-248.

[2] M. G. Moreno-Armenta, G. Soto. Stability and electronic structure of intrinsic and intercalated copper nitride alloys[J]. Solid State Sciences, 2008, 10(5): 573-579.

[3] R. Deshmukh, E. Tervoort, J. Kach, et al. Assembly of ultrasmall Cu_3N nanoparticles into three-dimensional porous monolithic aerogels[J]. Dalton Transactions, 2016, 45(29): 11616-11619.

[4] Z. Wang, X. Q. Cao, D. N. Liu, et al. Copper-Nitride Nanowires Array: An Efficient Dual-Functional Catalyst Electrode for Sensitive and Selective Non-Enzymatic Glucose and Hydrogen Peroxide Sensing[J]. Chemistry-a European Journal, 2017, 23(21): 4986-4989.

[5] C. Y. Su, B. H. Liu, T. J. Lin, et al. Carbon nanotube-supported Cu_3N nanocrystals as a highly active catalyst for oxygen reduction reaction[J]. Journal of Materials Chemistry A, 2015, 3(37): 18983-18990.

[6] A. Strozecka, J. C. Li, R. Schurmann, et al. Electroluminescence of copper-nitride nanocrystals[J]. Physical Review B, 2014, 90(19): 195420.

[7] J. R. Wang, F. Li, X. B. Liu, et al. Cu_3N and its analogs: a

new class of electrodes for lithium ion batteries[J]. Journal of Materials Chemistry A,2017,5(18):8762-8768.

[8] J. Timoshenko, A. Anspoks, A. Kalinko, et al. Thermal disorder and correlation effects in anti-perovskite-type copper nitride[J]. Acta Materialia,2017,129:61-71.

[9] G. G. Zhang, P. X. Yan, Z. G. Wu, et al. The effect of hydrogen on copper nitride thin films deposited by magnetron sputtering[J]. Applied Surface Science,2008,254(16):5012-5015.

[10] X. L. Tian, H. B. Tang, J. M. Luo, et al. High-Performance Core-Shell Catalyst with Nitride Nanoparticles as a Core: Well-Defined Titanium Copper Nitride Coated with an Atomic Pt Layer for the Oxygen Reduction Reaction[J]. Acs Catalysis, 2017,7(6):3810-3817.

[11] D. I. Bazhanov, O. V. Stepanyuk, O. V. Farberovich, et al. Classical and quantum aspects of spin interaction in 3d chains on a Cu_3N-Cu(110) molecular network[J]. Physical Review B, 2016,93(3):35444.

[12] K. Matsuzaki, T. Okazaki, Y. S. Lee, et al. Controlled bipolar doping in Cu_3N (100) thin films[J]. Applied Physics Letters,2014,105(22):35004.

[13] X. J. Li, A. L. Hector, J. R. Owen. Evaluation of Cu_3N and CuO as Negative Electrode Materials for Sodium Batteries[J]. Journal of Physical Chemistry C,2014,118(51):29568-29573.

[14] X. Y. Fan, Z. J. Li, A. L. Meng, et al. Study on the structure, morphology and properties of Fe-doped Cu_3N films[J]. Journal of Physics D-Applied Physics,2014,47(18):185304.

[15] V. C. Zoldan, R. Faccio, C. L. Gao, et al. Coupling of Cobalt-Tetraphenylporphyrin Molecules to a Copper Nitride Layer[J]. Journal of Physical Chemistry C, 2013, 117(31): 15984-15990.

[16] T. Nakamura, H. Hayashi, T. Ebina. Preparation of copper nitride nanoparticles using urea as a nitrogen source in a long-chain alcohol[J]. Journal of Nanoparticle Research,2014,16(11):2699.

[17] A. Daoudi, B. A. Touimi, J. P. Flament, et al. Potential Energy Curves and Electronic Structure of Copper Nitrides CuN and CuN^+ Versus CuO and CuO^+[J]. Journal of Molecular Spectroscopy,1999,194(1):8.

[18] G. H. Yue, P. X. Yan, J. Z. Liu, et al. Copper nitride thin film prepared by reactive radio-frequency magnetron sputtering[J]. Journal of Applied Physics,2005,98(10):890.

[19] J. F. Pierson, D. Horwat. Addition of silver in copper nitride films deposited by reactive magnetron sputtering[J]. Scripta Materialia,2008,58(7):568-570.

[20] T. Wang, X. J. Pan, X. M. Wang, et al. Field emission property of copper nitride thin film deposited by reactive magnetron sputtering[J]. Applied Surface Science,2008,254(21):6817-6819.

[21] L. X. Yang, J. G. Zhao, Y. Yu, et al. Metallization of Cu_3N semiconductor under high pressure[J]. Chinese Physics Letters,2006,23(2):426-427.

[22] J. R. Xiao, Y. W. Li, A. H. Jiang. Structure, Optical Property and Thermal Stability of Copper Nitride Films Prepared by Reactive Radio Frequency Magnetron Sputtering[J]. Journal of Materials Science & Technology,2011,27(5):403-407.

[23] W. Yu, J. G. Zhao, C. Q. Jin. Simultaneous softening of Cu_3N phonon modes along the T-2 line under pressure: A first-principles calculation[J]. Physical Review B,2005,72(21):214116.

[24] D. M. Borsa, S. Grachev, C. Presura, et al. Growth and properties of Cu_3N films and Cu_3N/gamma'-Fe_4N bilayers[J].

Applied Physics Letters,2002,80(10):1823-1825.

[25] J. Blucher, K. Bang, B. C. Giessen. Preparation of the metastable interstitial copper nitride, Cu_4N, by d. c. plasma ion nitriding[J]. Materials Science & Engineering A,1989,117(5): L1-L3.

[26] Z. G. Wu, W. W. Zhang, L. F. Bai, et al. Preparation and properties of nano-structure Cu_3N thin films[J]. Acta Physica Sinica,2005,54(4):1687-1692.

[27] M. Asano, K. Umeda, A. Tasaki. Cu_3N Thin-Film for a New Light Recording Media[J]. Japanese Journal of Applied Physics Part 1-Regular Papers Short Notes & Review Papers, 1990,29(10):1985-1986.

[28] T. Nosaka, M. Yoshitake, A. Okamoto, et al. Thermal decomposition of copper nitride thin films and dots formation by electron beam writing[J]. Applied Surface Science, 2001, 169: 358-361.

[29] D. M. Borsa, D. O. Boerma. Growth, structural and optical properties of Cu_3N films[J]. Surface Science, 2004, 548 (1-3):95-105.

[30] Q. A. Lu, X. Zhang, W. Zhu, et al. Reproducible resistive-switching behavior in copper-nitride thin film prepared by plasma-immersion ion implantation[J]. Physica Status Solidi a-Applications And Materials Science,2011,208(4):874-877.

[31] J. Wang, J. T. Chen, X. M. Yuan, et al. Copper nitride (Cu_3N) thin films deposited by RF magnetron sputtering[J]. Journal of Crystal Growth,2006,286(2):407-412.

[32] L. Maya. Deposition of Crystalline Binary Nitride Films of Tin, Copper, And Nickel by Reactive Sputtering[J]. Journal of Vacuum Science & Technology a-Vacuum Surfaces And Films, 1993,11(3):604-608.

[33] A. N. Fioretti, C. P. Schwartz, J. Vinson, et al. Understanding and control of bipolar self-doping in copper nitride[J]. Journal of Applied Physics, 2016, 119(18):32-41.

[34] M. G. Moreno-Armenta, A. Martinez-Ruiz, N. Takeuchi. Ab initio total energy calculations of copper nitride: the effect of lattice parameters and Cu content in the electronic properties [J]. Solid State Sciences, 2004, 6(1):9-14.

[35] G. H. Yue, P. X. Yan, J. Wang. Study on the preparation and properties of copper nitride thin films[J]. Journal of Crystal Growth, 2005, 274(3-4):464-468.

[36] N. Pereira, L. Dupont, J. M. Tarascon, et al. Electrochemistry of Cu_3N with lithium-A complex system with parallel processes[J]. Journal of the Electrochemical Society, 2003, 150(9):A1273-A1280.

[37] T. Nosaka, M. Yoshitake, A. Okamoto, et al. Copper nitride thin films prepared by reactive radio-frequency magnetron sputtering[J]. Thin Solid Films, 1999, 348(1-2):8-13.

[38] T. Wang, R. S. Li, X. J. Pan, et al. Improvement of Field Emission Characteristics of Copper Nitride Films with Increasing Copper Content [J]. Chinese Physics Letters, 2009, 26(6):236-238.

[39] R. Juza, H. Hahn Kupfernitrid. Metallamide und Metallnitride. VII[J]. Zeitschrift Für Anorganische Und Allgemeine Chemie, 1939, 241(2-3):172-178.

[40] J. M. Burkstrand, G. G. Kleiman, G. G. Tibbetts, et al. Study of the N-Cu(100) system[J]. Journal of Vacuum Science & Technology, 1976, 13(1):291-295.

[41] U. Zachwieja, H. Jacobs. ChemInform Abstract: Ammonothermal Synthesis of Copper Nitride, Cu_3N[J]. Cheminform, 1990, 21(35).

[42] S. Terada, H. Tanaka, K. Kubota. Heteroepitaxial growth of Cu_3N thin films[J]. Journal of Crystal Growth, 1989, 94(2): 567-568.

[43] T. Maruyama, T. Morishita. Copper nitride thin films prepared by radio-frequency reactive sputtering[J]. Journal of Applied Physics, 1995, 78(6): 4104-4107.

[44] J. R. Xiao, H. Xu, Y. F. Li, et al. Effect of nitrogen pressure on structure and optical band gap of copper nitride thin films[J]. Acta Physica Sinica, 2007, 56(7): 4169-4174.

[45] S. Ghosh, F. Singh, D. Choudhary, et al. Effect of substrate temperature on the physical properties of copper nitride films by r. f. reactive sputtering[J]. Surface & Coatings Technology, 2001, 142: 1034-1039.

[46] K. J. Kim, J. H. Kim, J. H. Kang. Structural and optical characterization of Cu_3N films prepared by reactive RF magnetron sputtering[J]. Journal of Crystal Growth, 2001, 222(4): 767-772.

[47] A. Majumdar, S. Drache, H. Wulff, et al. Strain Effects by Surface Oxidation of Cu_3N Thin Films Deposited by DC Magnetron Sputtering[J]. Coatings, 2017, 7(5): 64.

[48] Y. H. Zhao, Q. X. Zhang, S. J. Huang, et al. Effect of Magnetic Transition Metal (TM = V, Cr, and Mn) Dopant on Characteristics of Copper Nitride[J]. Journal of Superconductivity And Novel Magnetism, 2016, 29(9): 2351-2357.

[49] D. Dorranian, L. Dejam, A. H. Sari, et al. Structural and optical properties of copper nitride thin films in a reactive Ar/N-2 magnetron sputtering system[J]. European Physical Journal-Applied Physics, 2010, 50(2).

[50] F. Hadian, A. Rahmati, H. Movla, et al. Reactive DC magnetron sputter deposited copper nitride nano-crystalline thin

films:Growth and characterization[J]. Vacuum,2012,86(8): 1067-1072.

[51] N. Kaur, N. Choudhary, R. N. Goyal, et al. Magnetron sputtered Cu_3N/NiTiCu shape memory thin film heterostructures for MEMS applications[J]. Journal of Nanoparticle Research, 2013,15(3):1-16.

[52] X. A. Li, Z. L. Liu, A. Y. Zuo, et al. Properties of Al-doped copper nitride films prepared by reactive magnetron sputtering[J]. Journal of Wuhan University of Technology-Materials Science Edition,2007,22(3):446-449.

[53] A. Rahmati Ti-Containing. Cu_3N Nanostructure Thin Films:Experiment and Simulation on Reactive Magnetron Sputter-Assisted Nitridation[J]. Ieee Transactions on Plasma Science,2015,43(6):1969-1973.

[54] S. H. Zhang, Y. He, M. X. Li, et al. Synthesis and characterization of Cu_3N-WC nanocomposite films prepared by direct current magnetron sputtering[J]. Thin Solid Films, 2010, 518 (18):5227-5232.

[55] G. Soto, J. A. Diaz, W. de la Cruz. Copper nitride films produced by reactive pulsed laser deposition[J]. Materials Letters,2003,57(26-27):4130-4133.

[56] T. Torndahl, M. Ottosson, J. O. Carlsson. Growth of copper(I) nitride by ALD using copper(II) hexafluoroacetylacetonate,water,and ammonia as precursors[J]. Journal of the Electrochemical Society,2006,153(3):C146-C151.

[57] F. Fendrych, L. Soukup, L. Jastrabik, et al. Cu_3N films prepared by the low-pressure r. f. supersonic plasma jet reactor: Structure and optical properties[J]. Diamond And Related Materials,1999,8(8-9):1715-1719.

[58] L. Soukup, M. Sicha, F. Fendrych, et al. Copper nitride

thin films prepared by the RF plasma chemical reactor with low pressure supersonic single and multi-plasma jet system[J]. Surface & Coatings Technology,1999,116:321-326.

[59] X. Q. Du,Q. F. Zhou,Z. Yan,et al. The effects of oxygen plasma implantation on bipolar resistive-switching properties of copper nitride thin films[J]. Thin Solid Films, 2017, 625: 100-105.

[60] A. vonRichthofen,R. Domnick,R. Cremer. Cu—N films grown by reactive MSIP:Constitution,structure and morphology [J]. Mikrochimica Acta,1997,125(1-4):173-177.

[61] G. Sahoo, S. R. Meher, M. K. Jain. Room temperature growth of high crystalline quality Cu_3N thin films by modified activated reactive evaporation[J]. Materials Science And Engineering B-Advanced Functional Solid-State Materials, 2015, 191: 7-14.

[62] D. Wang, Y. Li. Controllable synthesis of Cu-based nanocrystals in ODA solvent[J]. Chemical Communications, 2011,47(12):3604.

[63] R. Szczesny, E. Szlyk, M. A. Wisniewski,et al. Facile preparation of copper nitride powders and nanostructured films [J]. Journal of Materials Chemistry C,2016,4(22):5031-5037.

[64] T. Nakamura, N. Hiyoshi, H. Hayashi, et al. Preparation of plate-like copper nitride nanoparticles from a fatty acid copper(II) salt and detailed observations by high resolution transmission electron microscopy and high-angle annular dark-field scanning transmission electron microscopy[J]. Materials Letters,2015,139:271-274.

[65] X. Y. Fan,Z. G. Wu,G. A. Zhang,et al. Ti-doped copper nitride films deposited by cylindrical magnetron sputtering[J]. Journal of Alloys And Compounds,2007,440(1-2):254-258.

[66] T. Nakamura, H. Hayashi, T. Hanaoka, et al. Preparation of Copper Nitride (Cu_3N) Nanoparticles in Long-Chain Alcohols at 130-200 degrees C and Nitridation Mechanism[J]. Inorganic Chemistry, 2014, 53(2):710-715.

[67] Y. Du, A. L. Ji, L. B. Ma, et al. Electrical conductivity and photoreflectance of nanocrystalline copper nitride thin films deposited at low temperature[J]. Journal of Crystal Growth, 2005, 280(3-4):490-494.

[68] J. R. Xiao, M. Qi, Y. Cheng, et al. Influences of nitrogen partial pressure on the optical properties of copper nitride films[J]. Rsc Advances, 2016, 6(47):40895-40899.

[69] J. R. Xiao, H. J. Shao, Y. W. Li, et al. Structure and Properties of the Copper Nitride Films Doped with Ti[J]. Integrated Ferroelectrics, 2012, 135:8-16.

[70] D. Fargue. Réductibilité des systèmes héréditaires à des systèmes dynamiques (régis par des équations différentielles ou aux dérivées partielles)[J]. C. R. Acad. Sci. Paris Sér. A-B, 1973: B471-B473.

[71] M. Mikula, D. Buc, E. Pincik. Electrical and optical properties of copper nitride thin films prepared by reactive DC magnetron sputtering[J]. Acta Physica Slovaca, 2001, 51(1): 35-43.

[72] A. Ji, C. Li, Y. Du, et al. Formation of a rosett nitride thin films via e pattern in copper nanocrystals gliding[J]. Nanotechnology, 2005, 16(10):2092-2095.

[73] U. Hahn, W. Weber. Electronic structure and chemical-bonding mechanism of Cu_3N, Cu_3NPd, and related Cu(I) compounds[J]. Physical Review B, 1996, 53(19):12684.

[74] Y. H. Zhao, J. Y. Zhao, T. Yang, et al. Enhanced write-once optical storage capacity of Cu_3N film by coupling with an

Al_2O_3 protective layer[J]. Ceramics International, 2016, 42(3): 4486-4490.

[75] S. Cho. Effect of substrate temperature on the properties of copper nitride thin films deposited by reactive magnetron sputtering[J]. Current Applied Physics, 2012, 12: S44-S47.

[76] N. Takeuchi. First-principles calculations of the ground-state properties and stability of ScN[J]. Physical Review B, 2002, 65(4): 045204.

[77] F. Gulo, A. Simon, J. Kohler, et al. Li-cu exchange in intercalated Cu_3N-With a remark on Cu_4N[J]. Angewandte Chemie-International Edition, 2004, 43(15): 2032-2034.

[78] Z. F. Hou. Effects of Cu, N, and Li intercalation on the structural stability and electronic structure of cubic Cu(3)N[J]. Solid State Sciences, 2008, 10(11): 1651-1657.

[79] A. Rahmati, H. Bidadi, K. Ahmadi, et al. Ti substituted nano-crystalline Cu_3N thin films[J]. Journal of Coatings Technology & Research, 2011, 8(2): 289-297.

[80] X. Y. Cui, A. Soon, A. E. Phillips, et al. First principles study of 3d transition metal doped Cu_3N[J]. Journal of Magnetism And Magnetic Materials, 2012, 324(19): 3138-3143.

[81] X. Y. Fan, Z. G. Wu, H. J. Li, et al. Morphology and thermal stability of Ti-doped copper nitride films[J]. Journal of Physics D-Applied Physics, 2007, 40(11): 3430-3435.

[82] A. L. Ji, N. P. Lu, L. Gao, et al. Electrical properties and thermal stability of Pd-doped copper nitride films[J]. Journal of Applied Physics, 2013, 113(4): 1985.

[83] J. A. Rodriguez, M. G. Moreno-Armenta, N. Takeuchi. Adsorption, diffusion, and incorporation of Pd in cubic (001) Cu_3N: A DFT study[J]. Journal of Alloys And Compounds, 2013, 576: 285-290.

[84] H. Chen, X. A. Li, J. Zhao, et al. First principles study on the influence of electronic configuration of M on Cu_3NM: M= Sc, Ti, V, Cr, Mn, Fe, Co, Ni[J]. Computational & Theoretical Chemistry, 2014, 1027(11): 33-38.

[85] L. Gao, A. L. Ji, W. B. Zhang, et al. Insertion of Zn atoms into Cu_3N lattice: Structural distortion and modification of electronic properties[J]. Journal of Crystal Growth, 2011, 321(1): 157-161.

[86] H. Y. Chen, X. A. Li, J. Y. Zhao, et al. First principles study of anti-ReO_3 type Cu_3N and Sc-doped Cu_3N on structural, elastic and electronic properties[J]. Computational And Theoretical Chemistry, 2013, 1018: 71-76.

[87] X. Y. Fan, Z. J. Li, A. Meng, et al. Improving the Thermal Stability of Cu_3N Films by Addition of Mn[J]. Journal of Materials Science & Technology, 2015, 31(8): 822-827.

[88] X. A. Li, J. P. Yang, A. Y. Zuo, et al. La-doped Copper Nitride Films Prepared by Reactive Magnetron Sputtering[J]. Journal of Materials Science & Technology, 2009, 25(2): 233-236.

[89] T. Ishikawa, M. Masuda, Y. Hayashi. Characterization on the electrical properties of copper nitride thin films and the effects of hydrogen implantation[J]. Journal of the Japan Institute of Metals, 1999, 63(5): 621-624.

[90] Y. Du, R. Huang, R. Song, et al. Effect of oxygen inclusion on microstructure and thermal stability of copper nitride thin films[J]. Journal of Materials Research, 2007, 22(11): 3052-3057.

[91] R. Cremer, M. Witthaut, D. Neuschutz, et al. Deposition and characterization of metastable Cu_3N layers for applications in optical data storage[J]. Mikrochimica Acta, 2000, 133(1-4): 299-302.

[92] J. F. Pierson Influence of bias voltage on copper nitride films deposited by reactive sputtering[J]. Surface Engineering, 2003,19(1):67-69.

[93] D. Y. Wang, N. Nakamine, Y. Hayashi. Properties of various sputter-deposited Cu—N thin films[J]. Journal of Vacuum Science & Technology a-Vacuum Surfaces And Films,1998, 16(4):2084-2092.

[94] J. L. Choi, E. G. Gillan. Solvothermal synthesis of nanocrystalline copper nitride from an energetically unstable copper azide precursor[J]. Inorganic Chemistry,2005,44(21):7385-7393.

[95] N. Kanoun-Bouayed, M. B. Kanoun, S. Goumri-Said. Structural stability, elastic constants, bonding characteristics and thermal properties of zincblende, rocksalt and fluorite phases in copper nitrides: plane-wave pseudo-potential ab initio calculations [J]. Central European Journal of Physics,2011,9(1):205-212.

[96] T. Maruyama, T. Morishita. Copper nitride and tin nitride thin films for write-once optical recording media[J]. Applied Physics Letters,1996,69(7):890-891.

[97] Z. Q. Liu, W. J. Wang, T. M. Wang, et al. Thermal stability of copper nitride films prepared by rf magnetron sputtering [J]. Thin Solid Films,1998,325(1-2):55-59.

[98] J. F. Pierson. Structure and properties of copper nitride films formed by reactive magnetron sputtering [J]. Vacuum, 2002,66(1):59-64.

[99] S. O. Chwa, K. H. Kim. Adhesion property of copper nitride film to silicon oxide substrate[J]. Journal of Materials Science Letters,1998,17(21):1835-1838.

[100] M. Ghoohestani, M. Karimipour, Z. Javdani. The effect of pressure on the physical properties of Cu_3N[J]. Physica Scripta,2014,89(3):240-240.

[101] S. Y. Wang, J. H. Qiu, X. Q. Wang, et al. The evolution of Cu_3N films irradiated by femtosecond laser pulses[J]. Applied Surface Science, 2013, 268: 387-390.

[102] Z. G. Ji, Y. H. Zhang, Y. Yuan, et al. Reactive DC magnetron deposition of copper nitride films for write-once optical recording[J]. Materials Letters, 2006, 60(29-30): 3758-3760.

[103] M. Birkett, C. N. Savory, A. N. Fioretti, et al. Atypically small temperature-dependence of the direct band gap in the metastable semiconductor copper nitride Cu_3N[J]. Physical Review B, 2017, 95(11): 115201.

[104] A. Ji, D. Yun, L. Gao, et al. Crystalline thin films of stoichiometric Cu_3N and intercalated Cu_3NM_x (M =   metals): Growth and physical properties[J]. Physica Status Solidi, 2010, 207(12): 2769-2780.

[105] C. M. Caskey, R. M. Richards, D. S. Ginley, et al. Thin film synthesis and properties of copper nitride, a metastable semiconductor[J]. Materials Horizons, 2014, 1(4): 424-430.

[106] N. Yamada, K. Maruya, Y. Yamaguchi, et al. p-to n-Type Conversion and Nonmetal-Metal Transition of Lithium-Inserted Cu_3N Films[J]. Chemistry of Materials, 2015, 27(23): 8076-8083.

[107] H. B. Wu, W. Chen. Copper Nitride Nanocubes: Size-Controlled Synthesis and Application as Cathode Catalyst in Alkaline Fuel Cells[J]. Journal of the American Chemical Society, 2011, 133(39): 15236-15239.

[108] W. Zhu, X. Zhang, X. N. Fu, et al. Resistive-switching behavior and mechanism in copper-nitride thin films prepared by DC magnetron sputtering[J]. Physica Status Solidi a-Applications And Materials Science, 2012, 209(10): 1996-2001.

[109] J. T. Luo. Effect of Additives on the Sintering of A-

morphous Nano-sized Silicon Nitride Powders[J]. Journal of Wuhan University of Technology-Materials Science Edition,2009,24(4): 537-539.

[110] D. écija, J. M. Gallego, R. Miranda. The adsorption of atomic N and the growth of copper nitrides on Cu(100)[J]. Surface Science,2009,603(15):2283-2289.

[111] S. Menzli, B. B. Hamada, I. Arbi, et al. Adsorption study of copper phthalocyanine on Si(111)($\sqrt{3} \times \sqrt{3}$)R30°Ag surface[J]. Applied Surface Science,2016,369:43-49.

[112] W. L. Wang, C. S. Yang. Silver nanoparticles embedded titania nanotube with tunable blue light band gap[J]. Materials Chemistry & Physics,2016,175:146-150.

[113] Y. Wu, J. Chen, H. Li, et al. Preparation and properties of Ag/DLC nanocomposite films fabricated by unbalanced magnetron sputtering[J]. Applied Surface Science, 2013, 284(11):165-170.

[114] T. Potlog, D. Duca, M. Dobromir. Temperature-dependent growth and XPS of Ag-doped ZnTe thin films deposited by close space sublimation method[J]. Applied Surface Science, 2015,352:33-37.

[115] F. Urbach. The Long-Wavelength Edge of Photographic Sensitivity and of the Electronic Absorption of Solids[J]. Physical Review,1953,92(5):1324-1324.

[116] A. B. Murphy. Band-gap determination from diffuse reflectance measurements of semiconductor films, and application to photoelectrochemical water-splitting[J]. Solar Energy Materials & Solar Cells,2007,91(14):1326-1337.

附 录

一、*Integrated Ferroelectrics*, 135:8-16, 2012

Structure and properties of the copper nitride thin films doped with Ti[①]

XIAO Jian-rong[1,2], JIANG Hao-yu[2], ZHOU Chang-rong[1]

1. Guangxi Key Laboratory of Information Materials, GuilinUniversity of Electronic Technology, Guilin 541004, China.

2. College of Science, GuilinUniversity of Technology, Guilin 541004, China

Abstract: Copper nitride (Cu_3N) thin films were prepared by a reactive radio frequency magnetron sputtering apparatus. The crystal structureof pure and Ti doped Cu_3N thin films were characterized by atomic force microscope (AFM) and X-ray diffraction (XRD) technique. The AFM images demonstrate that the pure Cu_3N thin films has a rough structure, while the thin films becomes compact after Ti doping. The grain size of the pure Cu_3N films is approximately 80nm and it becomes small when the

---

[①] Foundation Item: Project (11064003) supported by the National Science Foundation of China; Project (2010GXNSFA013122) supported bythe Guangxi Natural Science Foundation. Received date: 2011-03-28; Accepted date: 2011-Corresponding author. XIAO Jian-rong; Tel: +86-773-5891235; E-mail: csu_xiaojianrong@yahoo.com.cn

Ti is doped. The XRD tests indicate that the Ti atom has been doped in the Cu_3N thin films, and the lattice constantof the films becomes small after the Ti doped. The optical transmission spectrum and electrical resistivities were obtained by ultraviolet-visible (UV-VIS) spectrophotometer and four-probe method. The optical band gap and electrical resistivities of the thin films increase with the Ti content in the thin films.

Keywords: copper nitride thin films; Ti doped; structure; properties

1. Introduction

Metal nitrides(such as titanium nitride) and non-metallic nitrides (such as silicon nitride) show a widely variety of properties and applications[1]. Copper nitride(Cu_3N) thin films hasattracted considerable attention in recent years due toits applications for a new material of opticalstorage devices and microscopic metal links in integrated circuit fabrication processes[2-6]. Cu_3N has a cubic anti-ReO_3 type crystal structure with a lattice constant of 0.3815nm, and its density is 5.84g/cm^3[3,7]. Rather interesting, the Cuatoms do not occupy the close-packing sites of (111) plane. Therefore, the crystal structure has many vacantinterstitial sites that can be filled by anotheratom (such as plumbum, titanium) into the body center, then the electrical andoptical properties of the films may change remarkably[6,7]. Cu_3N thin films is stable in air at room temperature, and its decomposition temperature is approximately 300℃. Asano[8] reported that the Cu_3N thin films was decomposed into Cu thin films and N_2 by heating at 300℃ in argon, but Maruyama[3] reported that the decompose temperature of the Cu_3N thin films is 470℃. The resistivity of the thin films presents wide variation. Maruyama[9] reported Cu_3N thin films with a lattice constant above 0.3868nm are conductors, while the thin films with a lattice constant below 0.3868nm are insulators. Nosaka[10] reported that the resistivity of the thin films at room temperature is $\sim 10\Omega \cdot cm$ and it goes up to $\sim 1k\Omega \cdot cm$ for slightly N-rich films. According to theoretical calculations, the optical band gap of the thin films ranges from 0.13 to 0.9eV[11-13], while the experimental values ranges from 1.1 to 1.9eV[10,14,15]. Up to now, Cu_3N thin films have been successfully grown by different techniques, such as reactive radio frequency magnetronsputtering using copper and nitrogen as source materials[14,16,17], Cylindrical magnetron sputtering[7,18], DC sputtering[19], molecular beam epitaxy(MBE)[5], ion-assisted vapor deposition[9], reactive pulsed laser deposition in nitrogen ambient[20,21].

Copper nitride thin films has been studied for a long time, but most of the reported work on its structure, optical properties, thermal instability. Little attention has been paid to the doping thin films, which may bring on a remarkably change on the structure and properties of the Cu_3N thin films. The purpose of this work is to investigate the effect of Tidoping on the structure and properties of the Cu_3N thin films.

2. Experiments

2.1 Preparation of Samples

In this study, the copper nitridethinfilms samples were deposited by reactive radio frequency magnetron sputtering. The apparatus with $\Phi 450 \times 350 cm^3$ stainless-steel chamber was evacuated to $2.0 \times 10^{-3} Pa$ before admitting the nitrogen(N_2) and argon(Ar). The target was a copper (purity 99.99%) disc of 50mm diameter and 5mm thickness. The working gas was a mixture of N_2 and Ar at a total flow of 40 sccm. The Ti-doped content was controlled by the size of Ti sheet which adhered to the copper disc. The substrate was the glass sheet. Before loading into the chamber, the substrate was cleaned in an ultrasonic bath of acetone for ten minutes to remove residual organic contaminants. Then, it washed in deionized water and dried by blowing nitrogen gas. The radio frequency deposition power, sputtering pressure, and substrate temperature were fixed at 50W, 1.0Pa, and 50℃, respectively. The chamber vacuum just before growth was less than 1.0×10^{-3} Pa, and then argon was introduced for sputter cleaning in order to eliminate any impurity on the substrates (treatment parameters: 100W, 15min).

2.2 Sample Characterization

The crystalline phases of the Cu_3N thin films were characterized by an X-ray diffractometer (XRD, D/max 250) using Cu Ka radiation. The surface morphology of the thin films was investigated by anatomic force microscope (AFM, NT-MDT). The optical transmission spectrum was identified by an ultraviolet-visible spectrophotometer (UV-VIS, TU-1800) with the wavelength range of 300~1000nm. The indirect optical band gap of the thin films was obtained by extrapolating the absorption edge line to the abscissa as per the standard Tauc's plot technique[22,23]. The atomic percentage (at%) in the samples were investigated by an scanning fiber-optic microscope (SEM, Fei Quanta 200). The electrical resistivities of the samples were measured by four-probe method.

3. Results and Discussion

The atomic percentage(at%) of Cu, N, and Ti atoms in the films were obtained by SEM, and the lattice constant were calculated by the X-ray diffraction pattern. The lattice constant and the chemical compositions of the Cu_3N films before and after Ti doped are shown in Table 1.

Table 1 Chemical compositions and lattice constant of the films grown before and after Ti doped.

Sample	lattice constant/nm	Cu/at%	N/at%	Ti/at%
A	0.3831	85.37	14.63	0
B	0.3858	83.58	16.02	0.40
C	0.3863	81.72	17.85	1.63

Fig. 1 shows the XRD spectrum of pure and Ti-doped Cu_3N thin films. The XRD spectrum of the Cu_3N thin films shows that the Cu_3N crystallite has a typical anti-ReO_3 structure, and there is no Cu peaks in the spectrum. The results show that the product obtained under these conditions is not simple substance of copper. Sample B and C exhibit the Ti-doped effect on the structure of the Cu_3N thin films, and the films' growth prefers the $Cu_3N(100)$ direction at the high Ti content and the $Cu_3N(100)$ direction at low Ti content. It is obvious that two peaks of TiN appear in the spectrum, which associated with the following: TiN (111) at 36.7 degree and TiN (200) at 42.4 degree. It indicates that titanium element is not only doped effectively to the thin films, but also formed partly chemical bonds with N atoms. This result is not completely in good agreement with the report of the literature[7]. Compared with sample A and sample C, it can be found that the strongest peak of the pure thin films is Cu_3N (111), but the peak of the Cu_3N(111) is nearly equal to the Cu_3N (100) after Ti doping. It implies that the growth behavior of the Cu_3N thin films is affected by the doping of Ti, and the preferred orientation changes from (111) to (100). This is because the reactivity with nitrogen of Cu is lower than that of Ti, and it lead to the fact that the Ti atoms combined with the N atoms prevent the growth of Cu_3N thin films. Meanwhile, we found that the peak TiN (111) increases quickly than that of TiN (200) when the size of Ti sheet increases.

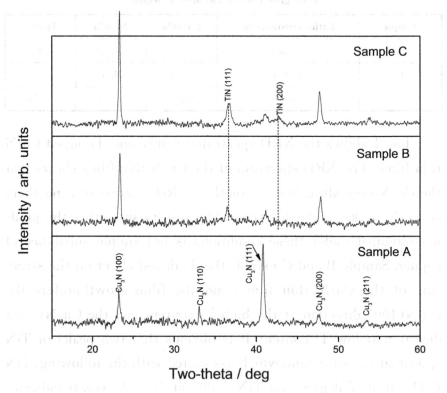

Fig. 1 thetype XRD spectra of the Cu_3N thin films

At the same time, the lattice constant and the grain size can been estimated from the X-ray diffraction pattern. Here, the grain size was obtained by the Debye-Scherre's formula:

$$D_0 = \frac{k\lambda}{\beta_0 \cos\theta}$$

Where D_0 is the grain size of the Cu_3N thin films, k is a constant, λ is the X-ray wavelength, β_0 is the full-width at half-maximum of the corresponding peak, and θ is the diffraction angle. The calculated results of the grain size of the Cu_3N thin films before Ti-doped is about 30nm, and this value is small than that we observed from the AFM image.

In order to investigatethe effect of Ti-doped on the surface morphology of the Cu_3N thin films, the AFM bidimensional (1000nm×1000nm) images of the Cu_3N thin films before and after doping are reported in Fig. 2. It can be seen that the Cu_3N thin films has tightly packed configuration and smooth surface morphology, and it is clear from the images that the surface of the thin films after doping reveals more compact structure. As shown in Fig. 2, the root mean square (RMS) surface roughness and the R_{max} (peak to peak) of the Cu_3N thin films is 3.603nm and 26.100nm, respectively. Whereas, after Ti doped, the RMS surface roughness of the thin films is 2.279nm and 15.447nm, respectively. Obviously, the observed morphology of the pure thin films has an average particle size of about 80nm. Whenas, the average particle size of the thin films is obviously reduced after Ti-doping and is about 40nm. Those observations suggest that the surface morphology of the Cu_3N thin films becomes smooth due to doping Ti. The variation of the surface morphology is due to: (1) the TiN is in face-centered cubic structure and has a lattice constant of 0.428nm, but the lattice constant of Cu_3N is 0.3815nm; (2) Ti doping changes the preferred orientation of the thin films. The pure thin films preferring to grow along the (111) orientation shows a pyramid-like morphology, while Ti-doping thin films preferring to grow along the (100) orientation shows a spherical-like morphology[7].

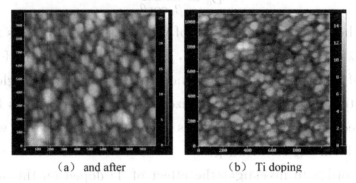

(a) and after　　　　　　(b) Ti doping

Fig. 2　AFM images of the Cu_3N thin films before

The UV-VIS transmittance spectra ofthe Cu_3N thin films on glass sheetsbefore and after Ti-doped are shown in Fig. 3. It can be seen that the high transmittance region is only in the range of 600~1000nm, and the UV absorption is very strong for the film before and after Ti-doping. But there is a little difference between them. The high transmittance region of the pure thin films almost kept a constant, while the transmittance region continuous increasewith the increasing wave number. Therefore, the high transmittance region shifts to high wavelength after the Ti doping.

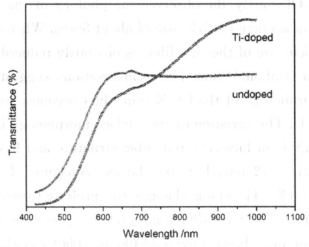

Fig. 3　The optical transmittance spectra ofthe Cu_3N thin films before and after Ti-doped.

From the UV-VIS transmission spectra, the absorption coefficient α of the Cu_3N thin films can be calculated from Tauc equation[22]: $(\alpha h v)^{1/2} = \beta(h v - E_g)$. The plot of $\alpha h v$ vs. $h v$ of the Cu_3N thin films prepared at before and after Ti doping is shown in Fig. 4. The intercept of the abscissa axis with the full line of the $(\alpha h v)^{1/2}$ vs. $h v$ plot allows the determination of optical band gap. As shown in Fig. 4, the optical band gap of the pure thin films is about 1.51eV, and the optical band gap of the Ti doped Cu_3N thin film is about 1.68eV. It is obvious that the optical band gap of Cu_3N thin films increases after the Ti doping. The reason of increasing optical band gap can be explained that some Ti atoms have not been combined with N atoms, which filledthe vacantinterstitial sites of the crystal structure of the Cu_3N thin films, thus lead to the increasing optical band gapof the Cu_3N thin films.

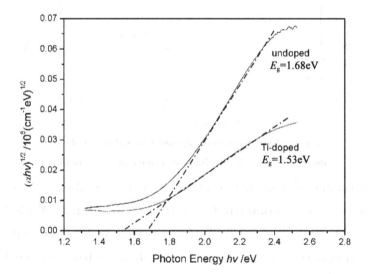

Fig. 4 The plot of $(\alpha h v)^{1/2}$ vs. $h v$ for estimation of optical band gap ofthe Cu_3N thin films before and after Ti-doped.

Fig. 5 shows the optical band gapand the electrical resistivity as functions of the content of Ti. It is obvious that the incorporation of Ti in the film changed the optical band gapand the electrical resistivity. The optical band gap of the thin films increases quickly with decreasing Ti content at first. Then, with the continued increasing Ti content, the optical band gap increased slowly. The increased optical band gap can be explained by the Burstein effect. Burstein thinks that the increase in the Fermi level in the conduction band of degenerate semiconductors leads to widening of the optical band gap.

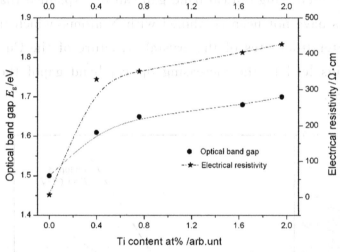

Fig. 5 The optical band gap andelectrical resistivity of the Ti-doped Cu_3N thin films as a function of Ti content.

In theory, the lattice constant of Cu_3N is about 0.3819nm, and the pure stoichiometric Cu_3N films is insulated. While it was found that the pure Cu_3N filmsdeposited byreactive radio frequency magnetron sputtering is a semiconductor and the lattice constant is 0.3831nm. As shown in Fig. 5 curve b, the electrical resistivity of the pure Cu_3N films is $7.52 \times 10^1 \Omega \cdot cm$. But the electrical resistivity increases rapidly to $3.29 \times 10^2 \Omega \cdot cm$ of the doped films with a few Ti content of 0.40at%, and then the electrical resistivity increases to $4.04×10^2 \Omega \cdot cm$ of the doped films with Ti content of 1.63at%. The electrical resistivity of the doped films increases with increasing the Ti content at first and then has no substantial variations. We think that the Ti in two kinds of films hasdifferent effects on the electrical resistivity, as the films with a few Ti content, the Ti atoms insert into the body of electrical lattice and it becomes a donor which releases a free electron as a carrier. Because of the increasing Ticontent in the films, a large quantity of Ti and N atoms can form the Ti-N bonds, which leads to the increase of the electrical resistivity. The changes of electrical resistivity suggest that the doped Ti atoms in the film do not contribute to dopants[7].

4. Conclusions

The pure and Ti-doped Cu_3N thin films were prepared by reactive radio frequency magnetron sputtering. The AFM images showthat the Cu_3N thin films became compact and uniform after the Ti doped. Since the doping Ti, the grain size of the Cu_3N thin films became small. The X-ray-diffraction results show that the thin films hasan anti-ReO_3 structure. The Ti-doped Cu_3N thin films'growthprefers the $Cu_3N(100)$ direction at the high Ti content and the $Cu_3N(100)$ direction at low Ti content. The UV-VIS transmittance spectra show that there is a good transmittance in the range of the visible light and infraredregion. The electrical resistivity of the Ti-doped Cu_3N films is a type semiconductor. The typical optical band gap of the Ti-doped Cu_3N thin films is 1.68eV, and it increases with the increasing Ti content in the thin films.

5. Acknowledgments

The authors would like to thank Dr. Gao Fei, Deng Chao-sheng, and Luo Cheng-lin for useful discussions and helps during the preparation of this work.

References

[1] H. Hu, X. L. Huang, M. M. Li, et al. Federated unscented particle filtering algorithm for SINS/CNS/GPS system [J]. Journal of Central South University Technology, 2010, 17: 778-785.

[2] D. Ecija, J. M. Gallgo, R. Miranda. The adsorption of atomic N and the growth of copper nitrides on Cu (100) [J]. Surface Science, 2009, 603: 2283-2289.

[3] T. Maruyama, T. Morishita. Copper nitride and tin nitride thin films for write-once optical recording media [J]. Applied Physics Letters, 1996, 69: 890-891.

[4] N. Matsunami, H. Kakiuchida, M. Tazawa, et al. Electronic and atomic structure modifications of copper nitride films by ion impact and phase separation [J]. Nuclear Instruments and Methods Physics Research B, 2009, 267: 2653-2656.

[5] D. M. Barsa, S. Grachev, C. Presura, et al. Growth and properties of Cu_3N films and $Cu_3N/\gamma-Fe_4N$ bilayers [J]. Applied Physics Letters, 2002, 80: 1823-1825.

[6] J. Wang, J. T. Chen, X. M. Yuan, et al. Copper nitride (Cu_3N) thin films deposited by RF magnetron sputtering [J]. Journal of Crystal Growth, 2006, 286: 407-412.

[7] X. Y. Fan, Z. G. Wu, G. A. Zhang, et al. Ti-doped copper

nitride films deposited by cylindrical magnetron sputtering [J]. Journal of Alloys & Compounds, 2007, 440: 254-258.

[8] M. Asano, K. Umeda, A. Tasaki. Cu_3N Thin Film for a New Light Recording Media [J]. Japanese Journal of Applied Physics, 1990, 29: 1985-1986.

[9] T. Maruyama, T. Morishita. Copper nitride thin films prepared by radio frequency reactive sputtering [J]. Journal of Applied Physics, 1995, 78: 4104-4107.

[10] T. Nosaka, M. Yoshitake, A. Okamato, et al. Copper nitride thin films prepared by reactive radio-frequency magnetron sputtering [J]. Thin Solid Films, 1999, 348: 8-13.

[11] Y. Wen, J. G. Zhao, C. Q. Jin. Simultaneous softening of Cu_3N phonon modes along the T_2 line under pressure: A first-principles calculation [J]. Physical Review B, 2005, 72: 214116.

[12] U. Hahn, W. Weber. Electronic structure and chemical-bonding mechanism of Cu_3N, Cu_3NPd, and related Cu(I) compounds [J]. Physical Review B, 1996, 53: 12684-12693.

[13] G. Moreno-armentam, A. Matinez-ruiza. Ab initio total energy calculations of copper nitride: the effect of lattice parameters and Cu content in the electronic properties [J]. Solid State Science, 2004, 6: 9-14.

[14] D. M. Borsa, D. O. Boerm. Growth, structural and optical properties of Cu_3N films [J]. Surface Science, 2004, 548: 95-105.

[15] F. Fendrych, L. Soulup, L. Jastrabik. Cu_3N films prepared by the low-pressure r. f. supersonic plasma jet reactor: Structure and optical properties [J]. Diamond and Related Materials, 1999, 8: 1715-1719.

[16] G. G. Zhang, P. X. Yan, Z. G. Wu, et al. The effect of hydrogen on copper nitride thin films deposited by magnetron

sputtering [J]. Applied Surface Science,2008,254:5012-5015.

[17] K. J. Kim,J. H. Kim,J. H. Kang. Structural and optical characterization of Cu_3N films prepared by reactive RF magnetron sputtering [J]. Journal of Crystal Growth, 2001, 222: 767-772.

[18] L. J. Cristina, R. A. Vidal. Surface characterization of nitride structures on Cu (001) formed by implantation of N ions: An AES, XPS and LEIS study [J]. Surface Science,2008,602: 3454-3458.

[19] N. Gordillo, R. Gonzalez-arrabal, M. S. Martin-Gonzalez,et al. DC triode sputtering deposition and characterization of N-rich copper nitride thin films: Role of chemical composition [J]. Journal of Crystal Growth,2008,310:4362-4367.

[20] S. Ghosh,F. Singh,D. Choudharyd,et al. Effect of substrate temperature on the physical properties of copper nitride films by r. f. reactive sputtering [J]. Surface and Coatings Technology,2001,142-144:1034-1309.

[21] G. Soto,J. A. Diaz,W. D. L. Cruz. Copper nitride films produced by reactive pulsed laser deposition [J]. Materials Letters,2003,57:4130-4133.

[22] J. Tauc, R. Grigorovici, A. Vancu. Amorphous and liquid semiconductors [J]. Physica Status Solidi,1966,15:627-630.

[23] J. F. Pierson. Structure and properties of copper nitride films formed by reactive magnetron sputtering [J]. Vacuum, 2002,66:59-64.

[24] G. H. Yue,P. X. Yan,J. Z. Liu,et al. Copper nitride thin film prepared by reactive radio-frequency magnetron sputtering [J]. Journal of Applied Physics,2005,98:103506.

二、*Journal of Materials Science & Technology*, 27(5): 403—407, 2011.

Structure, optical property and thermal stability of copper nitride films prepared by reactive radio frequency magnetron sputtering

Jianrong XIAO[1], Yanwei. LI[2], Aihua JIANG[1]

1. College of Science, Guilin University of Technology, Guilin 541004, China.

2. College of Chemistry and Bioengineering, Guilin University of Technology, Guilin, 541004, China

Copper nitride(Cu_3N) films were prepared by reactive radio frequency magnetron sputtering at various nitrogen partial pressures, and the films were annealed at different temperature. The crystal structure of the films was identified by X-ray diffraction technique. The Cu_3N films has a cubic anti-ReO_3 structure, and lattice constant is 0.3855 nm. With the increase of nitrogen partial pressure, the Cu_3N films are strongly textured with the crystal direction (100). The atomic force microscope images show that the films presence a smooth and compact morphology with nanocrystallites of about 70 nm in size. The films was further characterized by UV-visible spectrometer, and the optical band gap of the films was calculated from the Tauc equation. The typical value of optical band gap of the films is about 1.75 eV, and it increases with the increasing nitrogen partial pressure. The thermal property of the films was measured by a thermogravimetry, and

---

[1] Corresponding E-mail: csu_xiaojianrong@yahoo.com.cn

the decomposition temperature of the films was about 530K.

KEYWORDS: Cu_3N films; X-ray diffraction; structure; optical property; decomposition temperature

1. Introduction

In recent ten years, copper nitride(Cu_3N) has received considerable attention due to its interesting crystal structure and the rather low decomposition temperature, and it can be applied as new material for optical storage devices[1,2,3]. Since the Cu atoms do not occupy the close-packing sites of (111) plane, so the crystal structure has many vacant interstitial sites that can be filled by other atoms (such as plumbum) into the body center, which leads to remarkable change of the electrical and optical properties[4,5]. Maruyama[5] reported Cu_3N films with a lattice constant above 0.3868nm are conductors, while the films with a lattice constant below 0.3868nm are insulators. Yang[6] observed the gradual metallization of Cu_3N semiconductors by using high pressure, and the metallization pressure is 5.5GPa which is in good agreement with the first-principles calculations. Cu_3N has a dark green color and its density is $5.84g/cm^3$. Cu_3N is stable at room temperature, while the thermal decomposition occurs when the temperature reaches a certain value. Maruyama[1] and Asano[7] reported that the decomposition temperature of Cu_3N films is 743K and 573K, respectively. Zhang[8] reported that the Cu_3N films prepared at 10% H_2/N_2 ratios show poor stability and large weight gain compared to the Cu_3N films prepared at 0% H_2/N_2 ratios. Up to now, Cu_3N films have been successfully grown by various methods, such as magnetron sputtering using copper and nitrogen as source materials[5,9,10,11,12], molecular-beam epitaxy

(MBE)[2], ion-assisted vapor deposition[5], reactive pulsed laser deposition in nitrogen ambient and so on[13]. According to first-principles calculations, Cu_3N is a semiconductor with a rather low optical band gap (E_g) of ~0.13eV[3,14,15]. However, the experimental values ranging from 1.1 to 1.9eV were reported[10,16,17]. Wang reported that indirect optical band gap decreases with nitrogen flow rate increasing, and the typical value of E_g is 1.57eV[4]. Pierson found that the optical band gap of Cu_3N films ranges from 0.25 to 0.83eV, and it increases with increasing nitrogen partial pressure[18]. Gordillo et al investigated the influence of nitrogen excess on the optical response of N-rich Cu_3N films, and they found that the absorption spectra for the N-rich Cu_3N films were consistent with direct optical transitions corresponding to the stoichiometric semiconductor Cu_3N plus a free-carrier contribution that can be tuned in accordance with the N-excess[19].

Although the properties of copper nitride films have been studied for a long time, the structure, optical properties, and thermal instability are not fully understood. The purpose of this work is to investigate the effect of deposition parameter on the structure and optical properties of Cu_3N films and its thermal stability.

2. Experiment

2.1 Preparation of films

In preparing Cu_3N films, radio frequency reactive magnetron sputtering equipment was used. A pure copper (purity 99.99%) disc target of 50mm in diameter and 5mm thick was used for the magnetron sputtering. The working gas was a mixture of argon (Ar) and nitrogen (N_2). The total gas flow of nitrogen and argon was fixed at 60sccm, and the pressure was fixed at 0.8Pa. The nitrogen partial pressure (r, $r=[N_2]/\{[N_2]+[Ar]\}$) ranged from 0 to 1.0. The substrates were glass and quartz sheets. The substrate temperature (T) ranges from room temperature to 573K, and the discharge power (P_d) ranged from 50 to 300W. Before loading into the chamber, the substrates were cleaned in an ultrasonic bath of acetone and methanol for ten minutes to remove residual organic contaminants, respectively, and then washed in deionized water and dried by blowing nitrogen gas. The chamber vacuum just before growth was less than 1.0×10^{-3} Pa, and then argon was introduced for sputter cleaning in order to eliminate any impurity on the substrates (treatment parameters: 100W, 10min). The deposited samples were annealed in Ar atmosphere for 40min at temperature of 473K and 573K, respectively.

2.2 FilmsCharacterization

The crystalline phases of the films were analyzed by an X-ray diffractometer(XRD, D/max 250) using Cu Ka radiation, and the lattice constant of the films was obtained from XRD spectra. The surface morphology and chemical composition of the films were evaluated by atomic force microscope (AFM, NT-MDT). The optical band gap was identified by a UV-VIS spectrophotometer (TU-1800) with the wavelength range of 300~1000nm. The indirect optical band gap of the films was obtained by extrapolating the absorption edge line to the abscissa as per the standard Tauc's plot technique[20]. The thermal property of the films was measured by thermogravimetry (NETZSCH TG 209C).

3. Results and discussions

Fig. 1 shows the XRD spectra of Cu_3N films deposited at the nitrogen partial pressure of $r = 0.25, 0.50, 0.75$, and 1.0. The discharge power and substrate temperature during the sputtering were maintained at 100W and 393K, respectively.

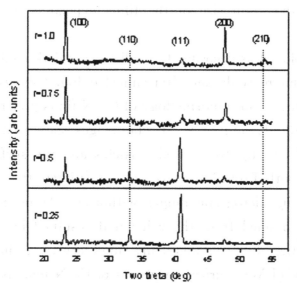

Fig. 1 XRD spectra of Cu_3N films deposited at different nitrogen partial pressure.

The XRD spectra of the films show that they are composed of Cu_3N crystallites with the anti-ReO_3 structure and the texture direction of the films changed from (111) to (100) with the increase of nitrogen partial pressure. As shown in Fig. 1, Along with the grains (111) there are other orientations such as (100) and (110) at the low nitrogen partial pressure of $r = 0.25$ and 0.75. In the XRD spectra of $r = 0.75$ and 1.0, the (100) peak is the strongest one, and the (111) peak is weaker than (200). It indicates that the nitrogen partial pressure in the sputtering gas mixture significantly affects the growth behavior.

In addition, the grain size and the lattice constant can also been estimated from the X-ray diffraction pattern of the films. Here, the grain size was obtained by the Debye-Scherre's formula[21]:

$$D_0 = \frac{k\lambda}{\beta_0 \cos\theta}$$

Where D_0 is the grain size of the films, β_0 is the full-width at half-maximum of the correspondingpeak, θ is the diffraction angle, λ is the X-ray wavelength, and k is a constant. The calculated results of the grain size of the films deposited at $r=0.5$ is about 27nm.

The color of as-deposited Cu_3N films was dark reddish-brown, and it became more darken after annealed. Fig. 2 shows the AFM images of the surface morphology of Cu_3N films prepared at $P=100W$, $T=300K$, and $r=0.50$. The images show that the Cu_3N films consist of tightly packed particles and the surface of the films is smooth. From Fig. 2(b), we can estimate that the size of crystallites in micrograph ranges within $50\sim80nm$, much more than that obtained from the calculated results. Obviously, each particle of the films contains many single crystal grains. Theroot mean square (RMS) surface roughness of Cu_3N films is 1.112nm, and the peak is 18.55nm. Meanwhile, we found that theRMS surface roughness of Cu_3N films increased with the increasing nitrogen partial pressures.

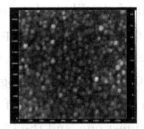

(a) three-dimensionalimage (b) two-dimensional image

Fig. 2 The AFM images ofCu₃N films

The UV-VIS transmittance spectra of the Cu_3N films on quartz crystal wafers prepared with various nitrogen partial pressures are shown in Fig. 3. Because the thickness of the films affects the transmission rate, the correspondingthicknesses of the films are marked in Fig. 3. It can be seen that the high transmittance region is only in the range of 600~1000nm and the UV absorption is very strong for the film deposited at any nitrogen partial pressure. With the increasing nitrogen partial pressure, the high transmittance region shifts to high wavelength and the transmission becomes weak.

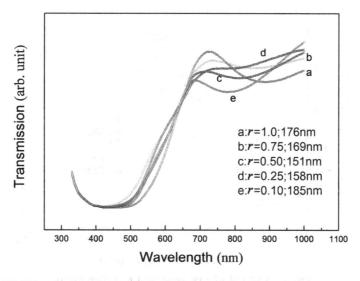

Fig. 3 UV-VIS transmission spectra of Cu_3N filmsprepared at various nitrogen partial pressure.

The reason is that the relative content of Cu—N bond increased with the increasing nitrogen partial pressure. From the UV-VIS transmission spectra, the absorption coefficient α of Cu_3N films can be calculated from Tauc equation[20]: $(\alpha h\upsilon)^{1/2} = \beta (h\upsilon - E_g)$. The plot of $\alpha h\upsilon$ vs. $h\upsilon$ of Cu_3N films prepared at various nitrogen partial pressures are shown in Fig. 4. The intercept of the abscissa axis with the full line of the $(\alpha h\upsilon)^{1/2}$ vs. $h\upsilon$ plot allows the determination of optical band gap. In our deposition conditions, the optical band gap of Cu_3N films ranges from 1.7 to 1.84eV. As shown in Fig. 4, the optical band gap E_g of Cu_3N films increases with the increasing nitrogen partial pressure. This can be explained that some Cu atoms have not been combined with N atoms in the films due to the lower nitrogen partial pressure, and lower E_g caused by the rich-Cu atoms in the films. These rich Cu atoms provide weakly local electron for Cu_3N films, resulting in a lower optical band gap. Higher E_g is caused by the phases of Cu_3N, and whole chemical band is Cu—N bond in the films.

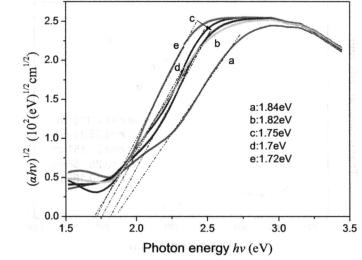

Fig. 4 A plot of $\alpha h\nu$ vs. $h\nu$ of the Cu_3N films grown at different nitrogen partial pressure.

Fig. 5 shows the lattice constants of the Cu_3N films as functions of different nitrogen partial pressure and optical band gap. It is clear that the lattice constant of the Cu_3N films increases quickly with decreasing nitrogen partial pressure. The optical band gap of the Cu_3N films increases with the increasing lattice constants. The dimension of optical band gap conforms that the films presents the electrical property of a semiconductor material. The changes of optical band gap may be due to the altered lattice constants of the films prepared conditions. At the same time, this fact suggests that the changes of the lattice constants of the films lead to the changes of the structure in the electronic states of the atoms, and bring to the changes of optical band gap of the films.

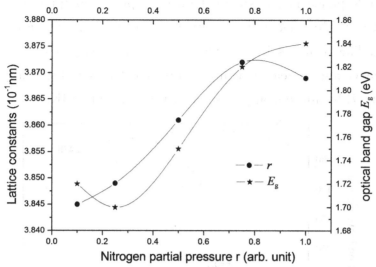

Fig. 5 Lattice constants of the Cu_3N films as a function of nitrogen partial pressure andoptical band gap

Fig. 6 shows the TG curves of the Cu_3N films prepared at nitrogen partial pressure of $r=0.1$ and $r=0.75$. The corresponding decomposition initiation temperature for the films prepared at $r=0.1$ and $r=0.75$ are 516K and 547K, respectively. This value is lower than those reported by Maruyama[1] and Nosaka[22]. In order to observe the thermal properties of the Cu_3N films, the samples of Cu_3N films were annealed in Ar atmosphere for 30min at temperatures of 473K and 573K, respectively.

The XRD spectra of the films before and after annealed are shown inFig. 7. It can be seen that the spectrum curve of the films before annealed has four peaks at (100),(110),(111) and (200) corresponded with copper nitride phase. The XRD spectra clearly exhibits (111)peak of Cu structure at 473K; meanwhile, the intensity of the (111) and (200) peaks of Cu_3N decreases a little. Obviously, the strongest peak is (111)peak of Cu structure at 573K, and the (200) peak of Cu structure appears. However, the peaks of (111) and (200) of Cu_3N almost disappear entirely. Cu_3N phase is transformed into Cu phase completely means that Cu_3N films decomposed entirely after annealing treatment at 573K. The result is in accordance with the analysis of TG. The TG curve and annealed results show that Cu_3N is thermally stable at 473K, and decomposes into Cu and N_2 at 516K.

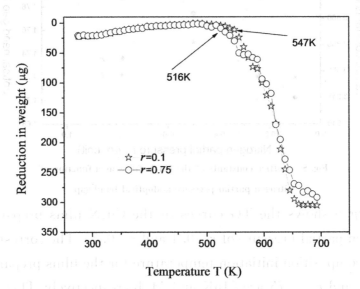

Fig. 6 Thermal analysis of Cu_3N films prepared at various nitrogen partial pressure.

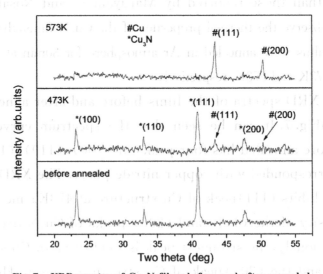

Fig. 7 XRD spectra of Cu_3N films before and after annealed.

4. Conclusions

Cu_3N films were prepared by reactive radio frequency magnetron sputtering at various nitrogen partial pressure and different discharge power. X-ray-diffraction results showed that the films has anti-ReO_3 structure. The lattice constant of the films is calculated from the XRD patterns. The typical lattice constant is 0.3855nm, and it increases with the increasing nitrogen partial pressure. The films has stronger texture along (100) direction when nitrogen flow rate increases. The Cu_3N films is thermally stable at 473K, and its decomposition temperature ranged from 516K to 547K. The optical band gap of the Cu_3N films depended on the nitrogen partial pressure, and increased with the increasing nitrogen partial pressure.

5. Acknowledgments

The project is supported by the National Science Foundation of China (Garnt No. 11064003) and the Guangxi NaturalScience Foundation (2010GXNSFA013122).

References

[1] T. Maruyama, T. Morishita. Copper nitride and tin nitride thin films for write-once optical recording media[J]. Applied Physics Letters, 1996, 69(7): 890-891.

[2] D. M. Borsa, S. Grachev, C. Presura, et al. Growth and properties of Cu_3N films and $Cu_3N/\gamma\text{-}Fe_4N$ bilayers[J]. Applied Physics Letters, 2002, 80(10): 1823-1825.

[3] U. Hahn, W. Weber. Electronic structure and chemical-bonding mechanism of Cu_3N, Cu_3NPd, and related $Cu(I)$ compounds[J]. Physics Review B,53 (1996) 12684.

[4] J. Wang, J. T. Chen, X. M. Yuan, et al. Copper nitride (Cu_3N) thin films deposited by RF magnetron sputtering[J]. Journal of Crystal Growth,2006,286:407-412.

[5] T. Maruyama, T. Morishita. Copper nitride thin films prepared by radio frequency reactive sputtering[J]. Journal of Applied Physics,1995,78(6):4104-4107.

[6] L. X. Yang, J. G. Zhao, Y. Yu, et al. Metallization for Cu_3N Semiconductor under High Pressure[J]. Chinese Physics Letters,2006,223:426-430.

[7] M. Asano, K. Umeda, A. Tasaki. Cu_3N Thin Film for a New Light Recording Media[J]. Japanese Jouenal of Applied Physics,1990,29:1985.

[8] G. G. Zhang, P. X. Yan, Z. G. Wu, et al. The effect of hydrogen on copper nitride thin films deposited by magnetron sputtering[J]. Applied Surface Science,2008,254:5012-5015.

[9] K. J. Kim, J. H. Kim, J. H. Kang. Structural and optical characterization of Cu_3N films prepared by reactive RF magnetron sputtering[J]. Journal of Crystal Growth,2001,222:767-772.

[10] T. Nosakaa, M. Yoshitake, A. Okamoto. Copper nitride thin films prepared by reactive radio-frequency magnetron sputtering [J]. Thin Solid Films,1999,348:8-13.

[11] S. Ghosh, F. Singh, D. Choudhary, et al. Effect of substrate temperature on the physical properties of copper nitride films by r. f. reactive sputtering[J]. Surface Coatings & Technology, 2001,142-144:1034-1039.

[12] J. F. Pierson, D. Horwat. Addition of silver in copper nitride films deposited by reactive magnetron sputtering[J].

Scripta Materials,2008,58:568-570.

[13] G. Soto, J. A. Diaz, W. de la Cruz. Copper nitride films produced by reactive pulsed laser deposition[J]. Materials Letters,2003,57:4130-4133.

[14] Y. Wen, J. G. Zhao, C. Q. Jin. Simultaneous softening of Cu_3N phonon modes along the T_2 line under pressure: A first-principles calculation [J]. Physics Review B,2005,72:214116.

[15] M. G. Moreno-Armenta, A. Martínez-Ruiz, N. Takeuchi. Ab initio total energy calculations of copper nitride: the effect of lattice parameters and Cu content in the electronic properties [J]. Solid State Science,2004,6:9-14.

[16] D. M. Borsa, D. O. Boerm. Growth, structural and optical properties of Cu_3N films[J]. Surface Scicence, 2004, 548: 95-105.

[17] F. Fendrych, L. Soukup, L. Jastrabik, et al. Cu_3N films prepared by the low-pressure r. f. supersonic plasma jet reactor: Structure and optical properties [J]. Diamond Related Materials, 1999,8:1715-1719.

[18] J. F. Pierson. Structure and properties of copper nitride films formed by reactive magnetron sputtering [J]. Vacuum, 2002,66:59-63.

[19] N. Gordillo, R. Gonzalez-Arrabal, M. S. Martin-Gonzalez, et al. DC triode sputtering deposition and characterization of N-rich coppernitride thin films : Role of chemical composition [J]. Journal of Crystal Growth,2008,310:4362-4367.

[20] J. Tauc, R. Grigorovici, A. Vancu. Amorphous and liquid semiconductors[J]. Physica Status Solidi,1966,15:627-634.

[21] G. H. Yue, P. X . Yan, J. Z. Liu, et al. Copper nitride thin film prepared by reactive radio-frequency magnetron sputtering[J]. Journal of Appllied Physics,2005,98:103506.

[22] T. Nosaka, M. Yoshitake, A. Okamoto, et al. Thermal decomposition of copper nitride thin films and dots formation by electron beam writing[J]. Applied Surface Science, 2001, 169-170:358-361.

三、*RSC Advances*, 6:40895—40899, 2016.

Influences of Nitrogen Partial Pressure on the Optical Properties of Copper Nitride Films

Jianrong Xiao[①], MengQi, Yong Cheng, Aihua Jiang[②], Yaping Zeng, Jiafeng Ma

College of Science, Guilin University of Technology, Guilin 541004, PR China

Abstract: Copper nitride (Cu_3N) films are made under different nitrogen pressuresviaradio frequency reaction and magnetron sputtering techniques. Scanning electron microscopy, X-ray diffraction, UV-visible spectrophotometry, and fluorescence spectrophotometry are conducted to test and analyze the structures and optical properties of the films. Films produced under distinct nitrogen pressures have compact surfaces and even similar particle sizes. In addition, Cu_3N (111) crystal faces are mainly formedin low nitrogen pressure, whereas Cu_3N (100) crystal faces are mainly generated in high nitrogen pressure. The photo-inducedluminous band of Cu_3N films is concentrated in the blue-violet light area. Moreover, the optical band gap E_g of films varies within the range of 1.23~1.91eV and increases along with R. Thistype of change is caused by the change of vacancy center inside the film

---

① Jianrong Xiao, Tel./fax: +86 773 3871615, E-mail address: xjr@glut.edu.cn
② Aihua Jiang, E-mail address: jah@glut.edu.cn

crystal and concentration of elementary substance copper atom causing a distinct defect energy level.

Keywords: Copper Nitride Films, Radio Frequency Reaction and Magnetron Sputtering, Nitrogen Pressure, Photoluminescence, Optical Band Gap

Copper, which is a cheap and environmentally friendly element, has an abundant natural resource. Cuprous compound is also an ideal luminous material. In recent years, photo-inducedlaminating research has achieved a series of successes[1,2]. Cu_3N crystal, which has become a researchfocus on semiconductor material, has also gained considerable attention from various researchers for its distinct structure and performance[3-7]. Cu_3N can replace disposal optical storage Te-based inorganic phase change material because of its simplicity and lack of toxicity[8,9]. Given its low decomposition temperature, Cu_3N can be used as buffer layer of integrated circuit, barrier layer of low magnetic resistance tunnel junction, and template of self-assembled materials[10]. It can also be used as a new battery material and catalyst additive material because of its unique chemical activity[11,12]. Cu_3N is also a competitor[13,14] of field emission material and has wide application prospect for its good electron emission property.

Cu_3N has anti-methyltrioxorhenium crystal structure. In the Cu_3N crystal, the s energy band of Cu atom can be overlapped with the p energy band of N atom, thereby forming a filled band. Thus, Cu_3N crystal is an insulator. Cu atom does not occupy a compact position of crystal lattice(111) in Cu_3N crystal, but leaves many interspaces in the crystal structure, hence making it easy for other atoms to fill in the center vacancy of Cu_3N crystal. The Cu_3N vacancy doping causes the change of crystal structure and leads Cu_3N to transform into a semiconductor or even conductor[15,16].

The application of Cu compoundas a phosphor material on OLED can satisfythe IQE 100% theoretically, which proves that it is an ideal luminous optical material. The photoluminescence efficiency of green light is close to 100%, whereas the photoluminescence efficiencies of blue and red lights are low. The workability of making OLED with most Cu compounds is also extremely low, and using traditional vacuum thermal evaporation or spin coating methods is difficult[17,18]. Many researchers reported about the optical performances of Cu_3N films. The Cu_3N E_g is calculated theoretically as $\sim 0.9eV$[19-21], whereas experimental results show that the value is within the range of $1.1 \sim 1.9eV$ and changes according to different production conditions[22,23]. Some production conditions, such as sputtering power, partial pressure, and doping in films, also affect the structure of the films and change the E_g to different extent[24-26]. In recent years, the demand for luminescent materials has been increasing. Therefore, research can adjust the luminous performance and determinethe luminous mechanismthrough improving the production process and adding different doping agents. Theelectroluminescence of Cu_3N nanocrystal hasalso been researched; Cu_3N nanocrystal can be used as a nanoscale light source[27]. However, no literature reports about the electroluminescence performance of Cu_3N films are available yet.

With its distinct structure, the chemical properties of Cu_3N are exceptional, but the relationship among the causes of performance change, change of crystal structure, and production process is still unknown. Thus, further research is necessary. In this paper, magnetron sputtering techniques are used to produce Cu_3N film. The crystal structure of Cu_3N films, photoluminescence spectrum with indoor temperature, change process of E_g, and micromechanism under different nitrogen partial pressures are also investigated.

1. Experiment

1.1 Sample Preparation

We designed the RF reaction magnetron sputtering equipment system, and it is manufactured by Shenyang Institute of Metal Research, Chinese Academy of Sciences. Cu (99.99%) is used as target, and the nitrogen (99.999%) and argon (99.999%) mixture is employed as working gas. (100)P silicon slice and quartz plate are used as substrate. The substrates were soaked in acetone and alcoholic solution individually, cleaned usingultrasonic wave for 20min, washed withdeionized water, and dried in a drying oven before placing them into the vacuum room. In every deposition, the background pressure in the vacuum room was lower than 1.0×10^{-3} Pa. Pre-sputtering in Ar(20sccm) environment was conducted for 10min before every sputtering. The surface and base of the target were cleaned, and nitrogen was pumped into deposit Cu_3N film samples. In sputtering, the total flow of nitrogen and argon was fixed at 40sccm; the sputtering pressure is 1.0Pa, and the frequency is 300W. In the experiment, $r(N_2/[N_2+Ar])$ was used as research parameter, where r can be 0.1, 0.2, 0.4, and 0.8. In deposition, the substrate temperature was fixed at 150℃, and the duration is 15min or 20min.

1.2 Sample Characterization

The surface of Cu_3N films and structure of film crystalswere characterized under the JSF-2100 field emissionscanning electron microscope and D/max 250 full-automatic X-ray diffractomer, respectively. The photoluminescence spectrum of films with indoor temperature was acquired using Cary Eclipse fluorescence spectrophotometer, and the exciting light was He-Cd laser with a wavelength of 190nm. Moreover, TU-1800 UV-visible spectrophotometer(UV-VIS) instrument was used to obtain the transmission spectrum of the film.

2. Results and Analysis

2.1 Film Surface

The color of the films made is closely relatedto containboth Cu_3N and Cu. The target used is pure copper, and the Cu_3N content becomes large with the increase of R. Therefore, film samples change their color from yellow of Cu to brownish black of Cu_3N.

Fig. 1a shows Cu_3N films deposited on the surface of a monocrystal silicon, which is compact and tight; the deposition power and nitrogen partial pressure(R) are 300W and 0.4, respectively. Fig. 1b shows the corresponding sectional view; the inner structure of the film samples is also compact, and the film thickness is around 500nm. Combining the deposition time and thickness of other samples, the average growth speed of Cu_3N films is about 33nm/min and decreases with increasing R. Fig. 2 shows EDS spectrum of Cu_3N films deposited with a R of 0.4. The figure also illustrates the detailed data of the apparent concentration of various elements and atomic mass percentages.

As observed from the EDS spectrum of Cu_3N films, the atomic mass percentage of Cu_3N films produced has distinct R (Table 1). Table 1 shows that the percentage of nitrogen atom mass first increases clearly with the increase of R, whereas the percentage of copper atom mass decreases. Second, the Cu_3N film contains highly concentrated O. Analysis shows that O mainly originated from two aspects: first, the oxygen content left at the bottom of vacuum room; and second, the water and oxygen contained in air, which is absorbed on the film surface before testing. In addition, the large Si content in films is caused by using monocrystalline silicone piece as deposition substrate.

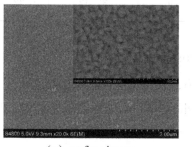 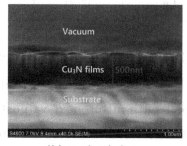

(a) surface image　　　　　(b) sectional view

Fig. 1　Scanning electron microscopyspectrum of Cu_3N films($R=0.4$)

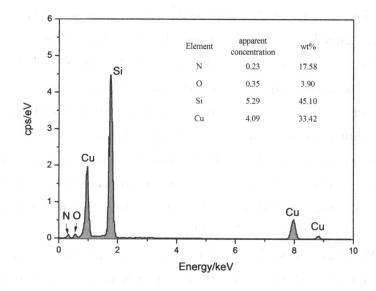

Fig. 2　EDS spectrum of Cu_3N films($R=0.2$)

· 143 ·

Table 1 EDS spectrum data of element (E)wt% of Cu_3N thin films deposited at different R

E \ R	0.1	0.2	0.4	0.8
N	14.94	16.30	17.58	27.91
O	1.65	2.96	3.90	1.91
Si	41.20	39.95	45.10	37.74
Cu	42.21	40.79	33.42	32.44

2.2 Analysis of Film Microstructure

Fig. 3 shows the X-ray diffraction (XRD) spectra of film samples with different R values (deposition power is 300W). With the increase of R, the diffraction peaks of Cu_3N films (100) and (110) at around 23° and 33°, respectively, are becoming increasingly stronger, and the change of (100) crystal face along with partial pressure is more evident than that of (110) crystal face. However, the diffraction peak of Cu_3N films (111) around 41° gradually becomes weaker, which means that films have preferred orientation growth with different R values; with low R, Cu_3N film grows on (111) crystal face, whereas on (100) and (110) crystal faces with high R. Thus, R can affect the law of Cu_3N film crystal preferred orientation growth[28]. The reason may be as follows: with low R, the absorbing nitrogen atom is inserted into the crystal lattice of copper atom, thereby forming a Cu—N chemical bond and films. The film also has the same growth law as that of Cu(111) and grows into Cu_3N (111) face. With high R, sufficient N atoms are present; these atoms can be combined with Cu on the target or substrate surface to form Cu—N. Subsequently, the films develop Cu_3N (100) and (110) surfaces based on the principle that the lowest crystal free energy has the preferential growth.

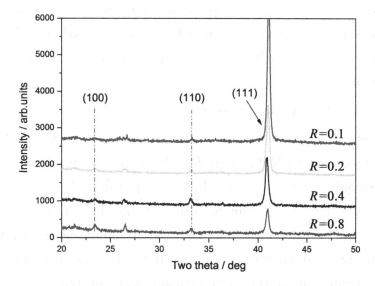

Fig. 3 X-ray diffractionspectra of Cu_3N films deposited by different R

Furthermore, with Braggs law (1) and cubic system interplanar distance law (2), the lattice constant of Cu_3N films can be acquired.

$$\lambda = 2d_{hkl}\sin\theta \qquad (1)$$

$$d_{hkl} = a\sqrt{h^2+k^2+l^2} \qquad (2)$$

where λ refers to the X-ray wavelength; θ refers to the diffraction angle; $h, k,$ and l refer to the crystal face constants; and d_{hkl} refers to the interplanar distance of the (hkl) crystal plane-family. The lattice constant a of Cu_3N films produced in distinct conditions can also be acquired(Table 2).

Table 2 Lattice constants(a) of Cu_3N films deposited at different R

R	0.8	0.4	0.2	0.10
a (Å)	3.831	3.845	3.858	3.872

The average lattice constant of Cu_3N films acquired from the experiment is around 3.851Å; the minimum and maximum values are 3.831Å and 3.872Å, respectively. The lattice constant increases along with the decrease of R. The value is also lower than the experimental values (around 3.868Å) obtained by other researchers, which may be caused by the distinct production processes of films.

2.3 Optical Transmission Properties of Film

Fig. 4a shows the UV-vistransmission spectrum of Cu_3N films produced with different R values. Cu_3N films have strong absorption properties in blue light and purple light regions. The films also have good transmission rate in the infrared region, which satisfies the character that photoluminescence spectrum has an excellent luminous zone in blue light and purplelight regions. According to the film transmission spectrum curve and based on the film optical constant law (3), which is calculated from the transmission spectroscopy, the absorption coefficient of films in weak, medium, and strong absorption regions can be calculated as follows:

$$\alpha = \ln(100/T)/d \tag{3}$$

where d refers to the film thickness, and T refers to light transmittance. Using the Tauc equation (4), the E_g of the film can be calculated as below[23]:

$$(\alpha h\nu)^{1/2} = A(h\nu - E_g) \tag{4}$$

where α, $h\nu$, and A refer to the absorption coefficient, photon energy, and constant, respectively. Using this equation, the E_g range of Cu_3N films made with distinct R is within 1.23 ~ 1.91eV, as shown in Fig. 4b. The E_g of Cu_3N film rapidly increases with the increase of R, and then the growth rapidly decreases; with the continuous increase of R, E_g also has a large increase. When R is 0.8 (lattice constant is 3.872nm), the E_g of the films increases rapidly to 1.91eV; this value is close to that reported in literature[29] that when the film lattice constant is larger than 3.868nm, it is a conductor, but an insulator when it is smaller than 3.868nm.

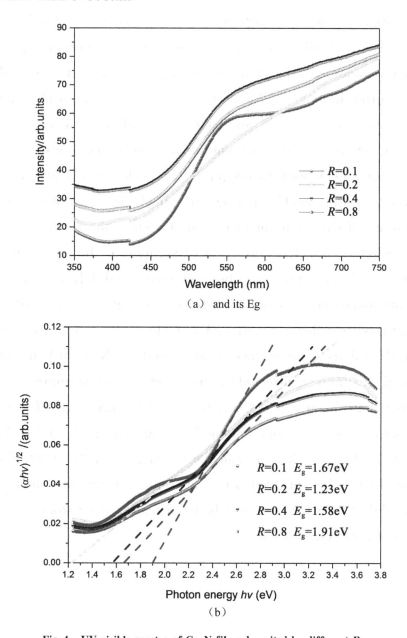

Fig. 4 UV-visible spectra of Cu_3N films deposited by different R

2.4 Luminescence Characteristics of Film Light

Fig. 5 shows the photoluminescence spectrum of Cu_3N films produced with different R values and indoor temperatures; their photoluminescence spectrum is mainly concentrated in the blue and purple light areas. For each film sample, three peaks of PL are observed at 421, 486, and 533nm. According to the changes of R, the intensity of distinct photoluminescence peak also varies. However, the homogeneous phases at peaks of 421 and 486nm are relatively large. Comparing with purple light and blue light emissions of Cu_3N, the main reasons for such observations are as follows: (1) defect energy level of vacancy in the center of the film and electron transition between compound defects; (2) Cu, as an interstitial atom, has the electron transition from defect energy level to valence bond; and (3) electron transition between interface defect and valence band around Cu_3N grain boundary; the E_g of film is co-affected by these factors. Furthermore, the intrinsic light band gap of Cu_3N film is within the range of 2.33~2.95eV, which is easily obtained through the peak of PL. This value is obviously higher than the E_g range of 1.23~1.91eV acquired from the Tauc equation. This finding means that the interface defect near Cu_3N crystal boundary is far away from the bottom of conduction band.

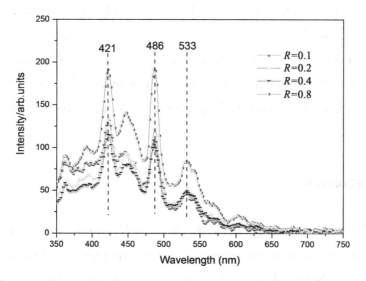

Fig. 5 Photoluminescencespectra of Cu_3N films deposited by different R

3. Analysis and Discussion

In our opinion, the aforementioned changes of E_g of Cu_3N films are mainly attributed to low R; the excessive sputtered elementary substance copper atom has not had a reaction with nitrogen at the surface of the target and bottom and subsequently deposits on the substrate. The formation of Cu_3N films is mainly through absorbing nitrogen atom to insert into the copper lattice, thereby forming a Cu—N bond. Simultaneously, some copper atoms deposited on the film surface have not formed a bond with nitrogen. Therefore, the central vacancy of Cu_3N lattice still contains many copper atoms, and the filling Cu atom provides weak localization electrons. For electrons combined on Cu_3N lattice with covalent bond, these weak localization electrons are non-localized, and their existence can change the electron density distribution inside the films. Thus, the film completely has the characteristics of a semiconductor and even a conductor, which has low E_g. When the R is large, the highly concentrated nitrogen atmosphere provides sufficient N atoms to react with Cu atoms and form Cu—N bond. Consequently, no Cu atoms fill in the gaps, and the E_g of the films is large. Therefore, the film is almost an insulator.

4. Conclusion

Cu_3N films are made with distinct R and RF frequency via RF reaction and magnetron sputtering techniques. The appearance of Cu_3N film, lattice structure, and optical characteristics are characterizedusing scanning electron microscopy, XRD, and UV-vis. The film surface is compact, and the particle sizesare on average. With the increase of R, the growth rate of Cu_3N (111) lattice is slightly decreased, whereas Cu_3N (100) and (110) lattices have relatively higher speed. The E_g of film samples is within the range of 1.23~1.91eV. With the increase of R, E_g first increases rapidly, andthen the growth speed decreases. Finally, the growth rate increases with the increase of R. This change is mainly caused by the central vacancy of films and non-bonded Cu atom causing change of defects, which further affects the electron density change inside the film lattice.

Conflict of Interests

The authors declare no conflict of interests regarding the publication of this paper.

5. Acknowledgments

The authors are grateful to the National Natural Science

Foundation of China (No. 11364011), Guangxi Natural Science Foundation (No. 2015GXNSFAA139004), and Innovation Project of Guangxi Graduate Education (No. YCSZ2015164) for their financial support.

References

[1] P. C. Ford, E. Cariati, J. Bourassa. Photoluminescence properties of multinuclear copper (I) compounds[J]. Chemical Reviews, 1999, 99, 3625-3647.

[2] Q. S. Zhang, T. Komino, S. P. Huang, et al. Triplet exciton confinement in green organic light-emitting diodes containing luminescent charge-transfer Cu (I) complexes [J]. Advanced Functional Materials, 2012, 22: 2327-2336.

[3] X. A. Li, Q. F. Bai, J. P. Yang, et al. Effect of N_2-gas flow rates on the structures and properties of copper nitride films prepared by reactive DC magnetron sputtering[J]. Vacuum, 2013, 89: 78-81.

[4] P. X. Xi, Z. H. Xu, D. Q. Gao, et al. Solvothermal synthesis of magnetic copper nitride nanocubes with highly electrocatalytic reduction properties[J]. RSC Advances, 2014, 4: 14206-14209.

[5] X. D. Xu, N. Y. Yuan, J. H. Qiu, et al. Formation of conductive copper lines by femtosecond laser irradiation of copper nitride film on plastic substrates[J]. Materials Research Bulletin, 2015, 65: 68-72.

[6] J. Park, K. Jin, B. Han, et al. Atomic layer deposition of copper nitride film and its application to copper seed layer for electrodeposition[J]. Thin Solid Films, 2014, 556: 434-439.

[7] X. Y. Fan, Z. J. Li, A. Meng, et al. Improving the Thermal Stability of Cu_3N Films by Addition of Mn[J]. Journal of

Materials Science & Technology,2015,31:822-827.

[8] T. Maruyama, T. Morishita. Copper nitride and tin nitride thin films for write-once optical recording media[J]. Applied Physics Letters,1996,69:890-891.

[9] Z. G. Ji, Y. H. Zhang, Y. Yuan. Reactive DC magnetron deposition of copper nitride films for write-once optical recording [J]. Materials Letters,2006,60:3758-3760.

[10] T. Wang, X. J. Pan, X. M. Wang, et al. Field emission property of copper nitride thin film deposited by reactive magnetron sputtering [J]. Applied Surface Science, 2008, 254: 6817-6819.

[11] A. S. Powell, R. I. Smith, D. H. Greogry, et al. Structure, stoichiometry and transport properties of lithium copper nitride battery materials: combined NMR and power neutron diffraction studies[J]. Physical Chemistry Chemical Physics, 2011, 13:10641-10647.

[12] B. S. Lee, M. Yi, S. Y. Chu, et al. Copper nitride nanoparticles supported on a superparamagnetic mesoporous microsphere for toxic-free click chemistry[J]. Chemical Communications,2010,46:3935-3937.

[13] A. L. Ji, C. R. Li, Y. Du, et al. Formation of a rosette pattern in copper nitride thin films via nanocrystals gliding[J]. Nanotechnology,2005,16:2092-2095.

[14] T. Wang, R. S. Li, X. J. Pan, et al. Improvement of Field Emission Characteristics of Copper Nitride Films with Increasing Copper Content [J]. Chinese Physics Letter, 2009, 26: 66801-66803.

[15] J. G. Zhao, S. J. You, L. X. Yang, et al. Structural phase transition of Cu_3N under high pressure[J]. Solid State Communications,2010,150:1521-1524.

[16] X. Y. Fan, Z. G. Wu, G. A. Zhang, et al. Ti-doped copper nitride films deposited by cylindrical magnetron sputtering [J]. Journal of Alloys & Compounds, 2007, 440: 254-258.

[17] R. Czerwieniec, J. Yu, H. Yersin. Blue-light emission of Cu(I) complexes and singlet harvesting[J]. Inorganic Chemistry, 2011, 50: 8293-8301.

[18] D. M. Zink, M. Bachle, T. Baumann, et al. Synthesis, structure, and characterization of dinuclear copper(I) halide complexes with P^N ligands featuring exciting photoluminescence properties[J]. Inorganic Chemistry, 2013(5), 522: 292-2305.

[19] Y. Wen, J. G. Zhao, C. Q. Jin. Simultaneous softening of Cu_3N phonon modes along the T2 line under pressure: A first-principles calculation [J]. Physics Review B, 2005, 72: 214116.

[20] M. G. Moreno-Armenta, A. Martínez-Ruiz. Ab initio total energy calculations of copper nitride: the effect of lattice parameters and Cu content in the electronic properties [J]. Solid State Science, 2004, 6: 9-14.

[21] U. Hahn, W. Weber. Electronic structure and chemical-bonding mechanism of Cu_3N, Cu_3NPd, and related Cu. I. compounds [J]. Physics Review B, 1996, 53: 12684-12693.

[22] D. M. Borsa, D. O. Boerm. Growth structural and optical properties of Cu_3N films [J]. Surface Science, 2004, 548: 95-105.

[23] F. Fendrych, L. Soukup, L. Jastrabik, et al. Cu_3N films prepared by the low-pressure r. f. supersonic plasma jet reactor: Structure and optical properties[J]. Diamond & Related Materials, 1999, 8: 1715-1719.

[24] J. R. Xiao, Y. W. Li, A. H. Jiang. Structure, optical property and thermal stability of copper nitride films prepared by reactive radio frequency magnetron sputtering[J]. Journal of Ma-

terials Science & Technology,2011,27:403-407.

[25] Z. G. Wu, W. W. Zhang, L. F. Bai, et al. Preparation and properties of nano-structure Cu_3N thin films[J]. Acta Physica Sinica,2005,54:1687-1692.

[26] G. M. Moreno-Armenta, G. Soto, N. Takeuchi. Ab initio calculations of non-stoichiometric copper nitride, pure and with palladium[J]. Journal of Alloys & Compounds,2011,17:1471-1476.

[27] A. Strozecka, J. C. Li, R. Schuermann, et al. Electroluminescence of copper-nitride nanocrystals[J]. Physics Review B, 2014,90:195420.

[28] L. X. Yang, J. G. Zhao, Y. Yu, et al. Metallization of Cu_3N Semiconductor under High Pressure[J]. Chinese Physics Letters,2006,23:426-427.

[29] T. Maruyama, T. Morishita. Copper nitride thin films prepared by radio frequency reactive sputtering[J]. Journal of Applied Physics,1995,78:4104-4107.

四、*Journal of Physics D: Applied Physics*, 2018, 51:055305

Crystal Structure and Optical Properties of Silver-doped Copper Nitride($Cu_3N:Ag$) Films Prepared by Magnetron Sputtering

Jianrong Xiao, Meng Qi, Chenyang Gong,

Zhiyong Wang, Aihua Jiang*, Jiafeng Ma, Yong Cheng*

College of Science, Guilin University of Technology, Guilin 541004, PR China

Corresponding Author: jah@glut.edu.cn (Aihua Jiang); hb_cy@163.com (Yong Cheng)

Silver-doped copper nitride($Cu_3N:Ag$) films were prepared

by using radio frequency and direct current magnetron sputtering method. The effects of silver doping on the crystal structure, surface morphology, and optical properties of the films were investigated. Results demonstrated that Ag doping influences the preferred growth orientation of the films. With the increased Ag content, the films grow from the (100) crystal face to the (111) crystal face. Ag doping changes the lattice constant, grain size, and optical band gap of $Cu_3N:Ag$ films. $Cu_3N:Ag$ films achieve the most remarkable crystallinity at 0.23 at% of Ag with lattice constant and optical band gap of 0.38188nm and 1.30eV, respectively. Moreover, the doped Ag atoms fill the hollows in the Cu_3N lattice and consequently reach the high-energy defect level. These atoms also generate impurity compound centers at high-energy level, shorten the non-equilibrium carrier lifetime of the films, and weaken the electronic transportation of the films.

Keywords: $Cu_3N:Ag$, magnetron sputtering, crystal structure, optical band gap

1. Introduction

Technological development results in high requirements on associated devices and materials. Film materials with broad band gaps and high transparencies are attracting increasing attention. Copper nitride (Cu_3N) with distinct structures and physical properties has attracted considerable attention as a candidate material of light storage device[1-5]. Cu_3N, a metastable covalent compound, possesses an antitrioxide lattice structure. The lattice constant of Cu_3N is 0.3815nm[6-9]. In Cu_3N crystals, Cu atoms fail to occupy the position of (111) crystal surface, thereby leaving many gaps in the crystal structure. If other metal atoms enter into these gaps, then the electrical and optical properties will change significantly[10-17]. Some scholars calculated the crystal energy band structure of Cu_3N by using the first principle, and they obtained an energy gap of approximately 1.0eV[14,18-21]. Many researchers prepared Cu_3N films by magnetron sputtering[1,9,11], reactive pulsed laser deposition[22,23], and molecular beam epitaxy[24,25] and obtained an optical band gap in the range of 1.1~2.3eV. The electroluminescence of Cu_3N crystals was also investigated; the Cu_3N crystal is an alternative material of nanoscale light source[26]. In addition, the photoluminescence (PL) characteristics of Cu_3N films have also been reported. The intrinsic luminescence band gap of Cu_3N films generally ranges between 2.33eV and 2.95eV[1]. Although many studies on optical properties of Cu_3N have been reported, the crystal structural changes caused by metal doping and preparation techniques and the mechanism of photoelectric property caused by crystal structural changes still remain unknown. Therefore, the Cu_3N properties should be further explored.

In the present study, silver-doped copper nitride($Cu_3N:Ag$) films were prepared by using radio frequency (RF) and direct current (DC) cosputtering. Influences of preparation techniques on $Cu_3N:Ag$ crystal structure are also discussed. Moreover, the mechanism of optical band gap and the defect state at a high energy level were analyzed.

2. Experiment

$Cu_3N:Ag$ films were prepared on a monocrystalline silicon and glass substrate by using JGP-450a magnetron sputtering. High-purity Cu and Ag (99.99%) were used as target materials. Nitrogen (99.99%) and argon (99.99%) were utilized as the source gas and working gas, respectively. The substrate was cleaned ultrasonically with acetone and alcohol and rinsed with deionized water. Finally, the substrate was dried in an oven under argon atmosphere. The ultimate pressure of the vacuum chamber was 1.0×10^{-3} Pa. Prior to deposition, the target material surface was sputter cleaned. The $Cu_3N:Ag$ films were prepared through simultaneous sputtering of RF and DC, and the RF power was maintained at 200W. The DC sputtering powers of Ag (P_w) were 0, 3, 6, and 9W, and corresponding samples were denoted as S_1, S_2, S_3, and S_4, respectively. The working pressure was maintained at 1.0Pa. The rotating speed of the substrate was 20rpm. The remaining parameters are listed in Table 1.

Table 1 Process parameters for the preparation of Ag doped Cu_3N films.

Parameters	Value
Ag target DC power (W)	0、3、6、9
Cu target RF power (W)	200
N_2 gas flow(sccm)	40
Ar gas flow(sccm)	10
Deposition temperature(℃)	RT
Deposition time(min)	15

The crystal structure of the films prepared on the silicon substrate was evaluated by an X-ray diffractometer (XRD, X'pert3 Powder). The Cu-targeted $K\alpha$ ray was used as diffraction source, and the X-ray optical wavelength was 0.15406nm. The surface and sectional morphologies of the films prepared on siliconwere observed by using a scanning electron microscope (SEM, JSF-2100), and the surface energy spectra of the films were obtained. The chemical composition of the film surface was analyzed using X-ray photoelectron spectroscopy (XPS, ESCAL-AB 250Xi). The PL characteristics of the films prepared on silicon substrate were investigated by using a Cary Eclipse fluorescence spectrophotometer with He-Cd laser. The wavelengths of the He-Cd laser were 370 and 310nm. The transmitted spectra of the films prepared on glass substratewere acquired by using an UV-visible spectrophotometer (UV-Vis, TU-1800).

3. Results and discussions

The SEM images of the Cu_3N and $Cu_3N:Ag$ films prepared on silicon are shown in Fig. 1. Crystal without Cu_3N shows considerably uniform grains and smooth, compact surfaces. Comparison ofthe images of $Cu_3N:Ag$ samples prepared under different P_w values indicated that grain size increases with the increased P_w value, and the film surface gradually becomes rough. When the P_w is 6W(Fig. 1c), sample displays large and regular crystal particles. In addition, the film surface is rough with high crystallinity. With the continuous increase in P_w, the film particles further shrink in size, and surface roughness decreases. This result may be due to that few doped Ag atoms scatter in Cu_3N films randomly under low P_w during the nucleation, growth, and film formation of Cu_3N films. This result influences the surface morphology of the films slightly. When the P_w is 6W, the proportion of doped Ag in the films increases to an appropriate concentration, which will affect the growth of Cu_3N films largely. The film particle size increases, and the particles become remarkably regular.

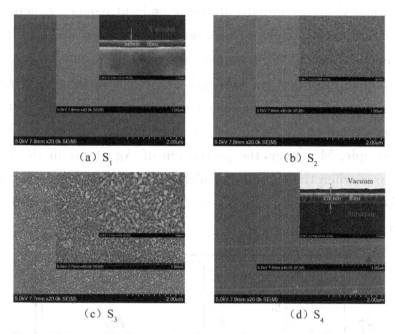

Fig. 1 High-magnification scanning electron microscope image of the surface morphology and cross-sections of the Cu_3N and $Cu_3N:Ag$ films

Table 2 EDS spectrum data of element (E) at% of the Cu_3N films and the $Cu_3N:Ag$ films. The value is the average of four test points, and the error bar is its root-mean-square deviation

E \ P_w	0	3W	6W	9W
Cu	28.34±0.88	27.04±0.82	31.30±1.00	28.85±1.16
N	19.05±0.87	18.28±0.62	22.74±0.94	20.06±0.72
O	11.72±1.43	7.34±1.78	6.67±0.57	6.21±0.36
Si	40.12±3.57	44.62±3.25	38.79±2.54	42.37±2.13
Ag	0	0.03±0.01	0.23±0.03	0.50±0.06

The EDS spectrum of $Cu_3N:Ag$ films (S_4) is shown in Fig. 2. The atom and mass fractions of elements in $Cu_3N:Ag$ films, which were prepared under P_w, can be obtained from the EDS test results (Table 2). First, the proportion of Ag atoms in the films increases with the increased P_w, whereas the proportions of Cu and C atoms initially decrease and subsequently increase. Second, many O atoms are observed on the films surface; these atoms mainly come from atmospheric and the residual oxygen in the vacuum chamber[27]. The high Si content in films is attributed to the monocrystalline silicon substrate; the silicon substrate was sputtered to the sample surface, at the time of preparing the test sample. Moreover, the proportion of Ag atoms in the films increases when the P_w is 9W, which indicated that the Ag content in the films increases quickly with the increased Ag sputtering power.

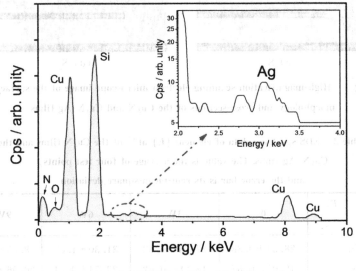

Fig. 2 EDS of Cu_3NAg films (S_4)

The XPS spectra of Cu_3N and Cu_3N:Ag films prepared on the silicon substrate are shown in Fig. 3. Elements in the films are completely consistent with the EDS results. Fig. 3 shows two evident adjacent peaks at 368.08 and 374.08eV, which correspond to $Ag3d_{5/2}$ and $Ag3d_{3/2}$, respectively[28,29]. The peak at 368.08eV reflects that Ag distributes in antitrioxide structured Cu_3N network as metal phase (Ag^0) and exists as Ag nanocluster[30]. Given that silver exists in Cu_3N films as Ag^0 and Ag^+, silver may play three roles in Cu_3N films: filling gap in Cu_3N structure, substituting Cu in Cu_3N crystals, and combining with O to form Ag_2O[31].

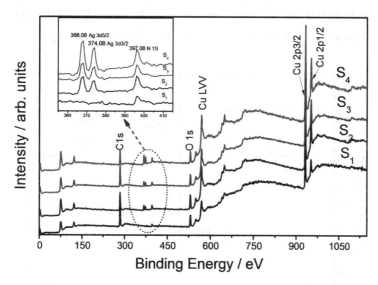

Fig. 3 X-ray photoelectron spectroscopy spectra of Cu_3N and Cu_3N:Ag films showing the Ag 3d peak

The XRD spectra of Cu_3N and Cu_3N:Ag films prepared on the silicon substrate are shown in Fig. 4. Peaks with the maximum intensity at (100), (111), and (200) conform to Cu_3N (PDF 47-1088). Few impurity peaks are also observed. These findings implied that the films possess high-quality and single-phase composition. Evidently, Cu_3N showsa preferred growth under different Ag concentrations: prior to Ag doping, the (100) crystal surface of Cu_3N crystal shows the strongest diffraction peak, and the (111) crystal surfacepresents weak peak position. With the increased Ag concentration in the films, the diffraction peak intensity of (100) and (200) crystal surfaces declines gradually, whereas that of (111) crystal surface increases remarkably. With further increased Ag concentration in the films, the diffraction peak intensity of (111) crystal surface reduces relatively again. This finding is similar with previously reported preferred growth of the films under changes in doping contents or preparation conditions[6,32-35]. Furthermore, the diffraction peak close to 56° is generated by the diffraction in the (220) crystal surface of Ag, which proved that Ag exist in the filmsas metal phase[17].

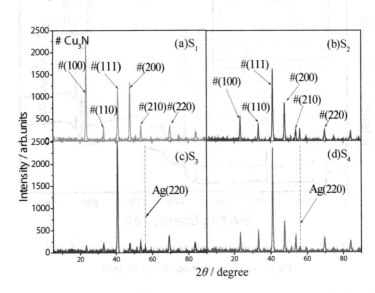

Fig. 4 X-ray diffractometer spectra of Cu_3N and $Cu_3N:Ag$ films

A strong diffraction peak, which is the small-angle (100) crystal surface, was used as reference. According to the Bragg's equation(1), cubic system distance equation (2), and Scherrer equation (3), the lattice constant and average grain size of Cu_3N:Ag crystal can be calculated as follows[1,34,36].

$$\lambda = 2d_{hkl}\sin\theta ; \quad (1)$$

$$d_{hkl} = a/\sqrt{h^2 + k^2 + l^2} ; \quad (2)$$

$$D = K\lambda/\beta\cos\theta ; \quad (3)$$

where h, k, and l are indexes of crystal face; a is a lattice constant; D is the average grain size; λ is the X-ray wavelength; θ is the diffraction angle; β is the full width at half maximum of the diffraction peak, K is Scherrer constant (0.89), and d_{hkl} is the interplanar distance of the (hkl) family of crystal surfaces. Therefore, the lattice constant and average grain size of the $Cu_3N:Ag$ crystal under different parameters are obtained (Table 3).

Table 3 The lattice constant (a) and the grain size (D) of Cu_3N and $Cu_3N:Ag$ filmscrystals prepared of different P_w

P_w (W)	0	3	6	9
a (nm)	0.38026	0.38029	0.38188	0.38180
D (nm)	1.6369	1.7824	1.8485	1.7788

Table 3 indicates that with the increased Ag content, the lattice constant of the $Cu_3N:Ag$ crystal shows an inverted V-shaped variation and reaches the maximum (0.38188nm) when P_w is 6W. This finding is different from reports that lattice constant decreases after partial doping[6,9]. The lattice constant of $Cu_3N:$ Ag prepared under the same conditions in this study is relatively smaller than that of mostly reported values[1,17,37,38]. This difference may be caused by the different experimental preparation techniques.

The transmitted spectra of Cu_3N and $Cu_3N:Ag$ films prepared on glass are shown in Fig. 5(a). The spectra change with P_w, and the film thickness is displayed. Evidently, the transmittance from UV to green light is almost 0. This finding exhibits differences with our previous research results[1,11,17,33]. Given that the $Cu_3N:Ag$ films are relatively thick, the transmittance is relatively low. The transmittance decreases remarkably after Ag doping. This result may be attributed to that after Ag doping; the film surface becomes rough and causesa diffused reflection of incident light, thereby influencing the transmittance of the films. At P_w of 6W, the Ag in films can reflect and absorb incident light strongly and exhibit high surface roughness. Consequently, the transmittance of $Cu_3N:Ag$ films is quickly reduced.

The Urbach energy (E_u) and optical band gap (E_g) of the films can be calculated by combining the transmittance spectra, Urbach model[39], and Tauc model[40,41]. The E_u reflects the changed rate of band tail density with energy index distribution, that is, the band tail width. E_u shows consistent changes with internal defect density in the films. High E_u means high internal defect density of the films. Small E_u means small internal defect density. The calculation formulas of E_u and E_g are as follows:

$$\alpha h\nu = A\exp(-h\nu/E_u); \quad (4)$$
$$(\alpha h\nu)^{1/2} = B(h\nu - E_g); \quad (5)$$

where A_0 and B are constant values, and $h\nu$ is the optical energy. The absorption coefficient α is calculated by the following equation:

$$\alpha = \ln[100/T]/d, \quad (6)$$

where T is the transmittance, and d is film thickness. The relationship between α and $h\nu$ of $Cu_3N:Ag$ films is shown in Fig. 5(b). E_u is the reciprocal of the slope of the linear section. The relationship between calculated $(\alpha h\nu)^{1/2}$ and $h\nu$ is shown in Fig. 5(c). E_g was obtained through linear extrapolation. Evidently, the minimum E_u of $Cu_3N:Ag$ films is achieved when the P_w is 6W; this finding indicated the good film crystallinity and the lowest internal defect density under this condition. This result also conforms to previous SEM results. E_g of $Cu_3N:Ag$ films decreases gradually with the increased P_w. E_g is 1.30eV when P_w is 6 W, as shown in Fig. 5(a). Therefore, the E_g of $Cu_3N:Ag$ films can be controlled reasonably by Ag doping concentration.

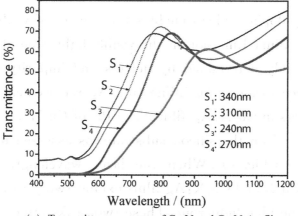
(a) Transmittance spectra of Cu$_3$N and Cu$_3$N:Ag films

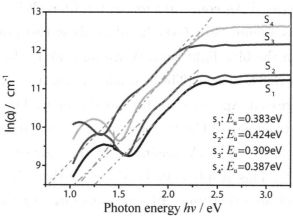
(b) relationship between α and $h\nu$ and determination of E_u

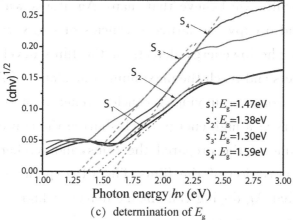
(c) determination of E_g

Fig. 5

The defect energy levels of Cu_3N films can be controlled effectively by doping, which enables the control of visible light emissions of films. Emission in the visible light zone of $Cu_3N:Ag$ films is mainly determined by defects and impurity concentrations. Lattice vacancy, Ag filling, and Ag replacement of Cu are major defects in $Cu_3N:Ag$ films. The PL of Cu_3N and $Cu_3N:Ag$ films prepared on the silicon substratewas assessed under room temperature (Fig. 6). When the laser wavelength is 310nm (Fig. 6a), Cu_3N and $Cu_3N:Ag$ films exhibit two peaks at 490 and 522nm in the blue light zone. These two peaks attenuate quickly with the increased Ag concentration in the films. When the laser wavelength is 370nm (Fig. 6b), the films show two peaks at 486 and 506nm in the blue light zone. A strong peak is also observed at 418nm in the UV zone. These peaks strengthen significantly with the increased Ag concentration in the films. Therefore, the intrinsic PL band gap of Cu_3N and $Cu_3N:Ag$ films is generally in the range of 2.38~2.97eV according to the light-emitting peak position of PL. This result is 0.81eV higher than the calculated E_g, which indicated that these PLs are generated by the high energy defects of the films. According to the luminescence spectra and EDS results, we believe that many Ag atoms will penetrate the films and occupy the lattice vacancies of Cu_3N with the increased P_w. The lowenergy defects of the films develop to high energy defects and reach the maximum at 0.23at% of Ag concentration. Ag atoms occupying vacancies generate impurity compound centers at a high energy level, shorten the non-equilibrium carrier lifetime of the prepared the films, and weaken the electronic transportation of the films. The PL experimental results confirmed that Ag doping can change defect at high energy level in Cu_3N films effectively.

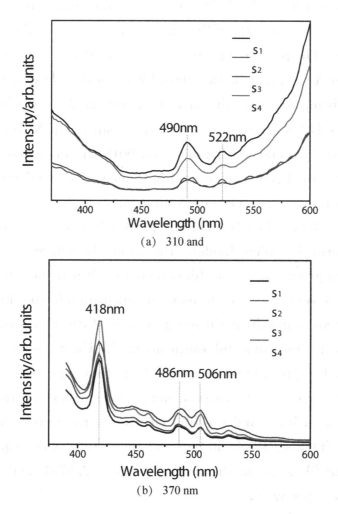

Fig. 6 Photoluminescence spectra of Cu_3N and $Cu_3N:Ag$ films

4. Conclusions

$Cu_3N:Ag$ films are prepared on glass and monocrystalline silicon substrates by RF and DC cosputtering. The effects of Ag content on crystal structure, energy level structure, and optical band gap of films are discussed. XRD results revealed that the diffraction peak of Ag (111) increases gradually with increased Ag content, whereas the diffraction peak intensity of Cu_3N (002) decreases. The preferred growth of Cu_3N (002) weakens with the increased Ag content in the films. SEM results showed that Ag content is important to the surface growth mode of Cu_3N films. When the DC sputtering power of Ag is small, Cu_3N films show smooth surface and small grain size. With the increased sputtering power of Ag, the Ag content in films increases, which results in rough film surface. Grain size first increases and subsequently decreases. When the silver content in the films is low, clusters are easily formed, which hinders the grain boundary migration. Thus, the grain size of the films increases. When the silver content in the film continues to increase, nucleation in the films rapidly increases, and the grain size decreases. With the increased Ag content, the Ag atoms fill vacancies in the Cu_3N lattice. Consequently, the optical band gap of the films first decreases and subsequently increases. Moreover, energy defects develop from low level to high level, and the film electronic transportation weakens. The intrinsic PL band gap and the optical band gap of the Cu_3N:Ag films are in the range of 2.38~2.97eV and 1.30~1.59eV, respectively.

Conflict of Interests

The authors declare noconflicts of interests with regard to the publication of this paper.

5. Acknowledgments

The authors are grateful to the National Natural Science Foundation of China(Grants No. 11364011), and Guangxi Natural Science Foundation (Grants No. 2017GXNSFAA 198121)

References

[1] J. R. Xiao, M. Qi, Y. Cheng, et al. Influences of nitrogen partial pressure on the optical properties of copper nitride films [J]. Rsc Advances, 2016, 6(47):40895-40899.

[2] W. Yu, J. G. Zhao, C. Q. Jin. Simultaneous softening of Cu_3N phonon modes along the T-2 line under pressure: A first-principles calculation[J]. Phys. Rev. B, 2005, 72(21):214116.

[3] Q. A. Lu, X. Zhang, W. Zhu, et al. Reproducible resistive-switching behavior in copper-nitride thin film prepared by plasma-immersion ion implantation[J]. Physica Status Solidi a-Applications And Materials Science, 2011, 208(4):874-877.

[4] A. Majumdar, S. Drache, H. Wulff, et al. Strain Effects by Surface Oxidation of Cu_3N Thin Films Deposited by DC Magnetron Sputtering[J]. Coatings, 2017, 7(5):64.

[5] G. Sahoo, M. K. Jain. Formation of CuO on thermal and laser-induced oxidation of Cu_3N thin films prepared by modified activated reactive evaporation[J]. Appl. Phys. A-mater., 2015, 118(3):1059-1066.

[6] Y. H. Zhao, Q. X. Zhang, S. J. Huang, et al. Effect of Magnetic Transition Metal(TM = V, Cr, and Mn) Dopant on Characteristics of Copper Nitride[J]. Journal of Superconductivity And Novel Magnetism, 2016, 29(9):2351-2357.

[7] X. J. Li, A. L. Hector, J. R. Owen. Evaluation of Cu_3N and CuO as Negative Electrode Materials for Sodium Batteries [J]. Journal of Physical Chemistry C, 2014, 118(51):29568-29573.

[8] R. Gonzalez-Arrabal, N. Gordillo, M. S. Martin-Gonzalez, et al. Thermal stability of copper nitride thin films: The role of

nitrogen migration[J]. J. Appl. Phys. ,2010,107(10):567.

[9] A. A. Yu, Y. H. Ma, A. S. Chen, et al. Thermal stability and optical properties of Sc-doped copper nitride films[J]. Vacuum,2017,141:243-248.

[10] N. Kaur, N. Choudhary, R. N. Goyal, et al. Magnetron sputtered Cu_3N/NiTiCu shape memory thin film heterostructures for MEMS applications[J]. Journal of Nanoparticle Research, 2013,15(3):1-16.

[11] J. R. Xiao, Y. W. Li, A. H. Jiang. Structure, Optical Property and Thermal Stability of Copper Nitride Films Prepared by Reactive Radio Frequency Magnetron Sputtering[J]. J. Mater. Sci. Technol. ,2011,27(5):403-407.

[12] A. L. Ji, N. P. Lu, L. Gao, et al. Electrical properties and thermal stability of Pd-doped copper nitride films[J]. J. Appl. Phys. ,2013,113(4):1985.

[13] X. Y. Fan, Z. G. Wu, H. J. Li, et al. Morphology and thermal stability of Ti-doped copper nitride films[J]. J. Phys. D. Appl. Phys. ,2007,40(11):3430-3435.

[14] U. Hahn. W. Weber Electronic structure and chemical-bonding mechanism of Cu_3N, Cu_3NPd, and related Cu(I) compounds[J]. Phys. Rev. B,1996,53(19):12684.

[15] X. Y. Fan, Z. G. Wu, G. A. Zhang, et al. Ti-doped copper nitride films deposited by cylindrical magnetron sputtering [J]. J. Alloy. Compd. ,2007,440(1-2):254-258.

[16] U. Zachwieja, H. Jacobs. Ammonothermalsynthese von kupfernitrid, Cu_3N[J]. Journal of The Less-Common Metals, 1990,161(1):175-184.

[17] J. F. Pierson, D. Horwat. Addition of silver in copper nitride films deposited by reactive magnetron sputtering[J]. Scripta Mater. ,2008,58(7):568-570.

[18] H. Chen, X. A. Li, J. Zhao, et al. First principles study on the influence of electronic configuration of M on Cu_3N M: M=Sc, Ti, V, Cr, Mn, Fe, Co, Ni[J]. Computational & Theoretical Chemistry, 2014, 1027(11): 33-38.

[19] J. A. Rodriguez, M. G. Moreno-Armenta, N. Takeuchi. Adsorption, diffusion, and incorporation of Pd in cubic (001) Cu_3N: A DFT study[J]. J. Alloy. Compd., 2013, 576: 285-290.

[20] M. G. Moreno-Armenta, G. Soto, N. Takeuchi. Ab initio calculations of non-stoichiometric copper nitride, pure and with palladium[J]. J. Alloy. Compd., 2011, 509(5): 1471-1476.

[21] M. G. Moreno-Armenta, A. Martinez-Ruiz, N. Takeuchi. Ab initio total energy calculations of copper nitride: the effect of lattice parameters and Cu content in the electronic properties [J]. Solid State Sci., 2004, 6(1): 9-14.

[22] G. Soto, J. A. Diaz, W. de la Cruz. Copper nitride films produced by reactive pulsed laser deposition[J]. Mater. Lett., 2003, 57(26-27): 4130-4133.

[23] T. Torndahl, M. Ottosson, J. O. Carlsson. Growth of copper(I) nitride by ALD using copper(II) hexafluoroacetylacetonate, water, and ammonia as precursors[J]. J. Electrochem. Soc., 2006, 153(3): C146-C151.

[24] D. M. Borsa, S. Grachev, C. Presura, et al. Growth and properties of Cu_3N films and Cu_3N/gamma'-Fe_4N bilayers[J]. Appl. Phys. Lett., 2002, 80(10): 1823-1825.

[25] S. Terada, H. Tanaka, K. Kubota. Heteroepitaxial growth of Cu_3N thin films[J]. J. Cryst. Growth, 1989, 94(2): 567-568.

[26] A. Strozecka, J. C. Li, R. Schurmann, et al. Electroluminescence of copper-nitridenanocrystals[J]. Phys. Rev. B, 2014, 90 (19): 195420.

[27] M. Mikula, D. Buc, E. Pincik. Electrical and optical properties of copper nitride thin films prepared by reactive DC magnetron sputtering[J]. Acta Phys. Slovaca., 2001, 51(1): 35-43.

[28] S. Menzli, B. B. Hamada, I. Arbi, et al. Adsorption study of copper phthalocyanine on Si(111)($\sqrt{3} \times \sqrt{3}$)R30°Ag surface[J]. Appl. Surf. Sci., 2016, 369: 43-49.

[29] W. L. Wang, C. S. Yang. Silver nanoparticles embedded titania nanotube with tunable blue light band gap[J]. Mater. Chem. Phys., 2016, 175: 146-150.

[30] Y. Wu, J. Chen, H. Li, et al. Preparation and properties of Ag/DLC nanocomposite films fabricated by unbalanced magnetron sputtering[J]. Appl. Surf. Sci., 2013, 284(11): 165-170.

[31] T. Potlog, D. Duca, M. Dobromir. Temperature-dependent growth and XPS of Ag-doped ZnTe thin films deposited by close space sublimation method[J]. Appl. Surf. Sci., 2015, 352: 33-37.

[32] D. Wang, N. Nakamine, Y. Hayashi. Properties of various sputter-deposited Cu—N thin films[J]. Journal of Vacuum Science & Technology A Vacuum Surfaces & Films, 1998, 16(4): 2084-2092.

[33] J. R. Xiao, H. Xu, Y. F. Li, et al. Effect of nitrogen pressure on structure and optical band gap of copper nitride thin films[J]. Acta Phys. Sin-ch. Ed., 2007, 56(7): 4169-4174.

[34] G. H. Yue, P. X. Yan, J. Wang. Study on the preparation and properties of copper nitride thin films[J]. J. Cryst. Growth, 2005, 274(3-4): 464-468.

[35] J. R. Xiao, H. J. Shao, Y. W. Li, et al. Structure and Properties of the Copper Nitride Films Doped with Ti[J]. Integr. Ferroelectr., 2012, 135: 8-16.

[36] M. Freeda, T. D. Subash. Comparision of Photoluminescence studies of Lanthanum, Terbium doped Calcium Aluminate nanophosphors($CaAl_2O_4$:La, $CaAl_2O_4$:Tb) by sol-gel method[J]. Mater. Today Proc., 2017, 4(2): 4302-4307.

[37] T. Maruyama, T. Morishita. Copper nitride thin films prepared by radio-frequency reactive sputtering[J]. J. Appl. Phys., 1995, 78(6): 4104-4107.

[38] J. Wang, J. T. Chen, X. M. Yuan, et al. Copper nitride (Cu_3N) thin films deposited by RF magnetron sputtering[J]. J. Cryst. Growth, 2006, 286(2): 407-412.

[39] F. Urbach. The Long-Wavelength Edge of Photographic Sensitivity and of the Electronic Absorption of Solids[J]. Physical Review, 1953, 92(5): 1324-1324.

[40] A. B. Murphy. Band-gap determination from diffuse reflectance measurements of semiconductor films, and application to photoelectrochemical water-splitting[J]. Solar Energy Materials & Solar Cells, 2007, 91(14): 1326-1337.

[41] F. Fendrych, L. Soukup, L. Jastrabik, et al. Cu_3N films prepared by the low-pressure r. f. supersonic plasma jet reactor: Structure and optical properties[J]. Diam. Relat. Mater., 1999, 8(8-9): 1715-1719.